The Digital Pencil

One-to-One Computing for Children

The
Digital
Pencil

One-to-One Computing
for Children

Jing Lei • Paul F. Conway • Yong Zhao

LEA Lawrence Erlbaum Associates
Taylor & Francis Group

New York   London

Lawrence Erlbaum Associates
Taylor & Francis Group
270 Madison Avenue
New York, NY 10016

Lawrence Erlbaum Associates
Taylor & Francis Group
2 Park Square
Milton Park, Abingdon
Oxon OX14 4RN

© 2008 by Taylor & Francis Group, LLC
Lawrence Erlbaum Associates is an imprint of Taylor & Francis Group, an Informa business

Printed in the United States of America on acid-free paper
10 9 8 7 6 5 4 3 2 1

International Standard Book Number-13: 978-0-8058-6060-3 (Hardcover)

Library of Congress Cataloging-in-Publication Data

Lei, Jing.
    The digital pencil : one-to-one computing for children / Jing Lei, Paul F. Conway, and Yong Zhao.
        p. cm.
    Includes bibliographical references and index.
    ISBN-13: 978-0-8058-6060-3 (alk. paper)
    ISBN-10: 0-8058-6060-6 (alk. paper)
    1. Education--Data processing. 2. Internet in education. 3. Laptop computers.
4. Computers and children. 5. Computer literacy. I. Conway, Paul F. II. Zhao, Yong, 1965- III. Title.

LB1028.43.L45 2008
372.133'4--dc22

    2007013850

Visit the Taylor & Francis Web site at
http://www.taylorandfrancis.com

Dedication

To our respective parents in Henan (China),
Laois (Ireland), and Sichuan (China)

Contents

Preface

The "Digital Pencil": An Introduction

Imagine that writing has just been invented in Foobar, a country that has managed to develop a highly sophisticated culture of poetry, philosophy and science using entirely oral means of expression. It occurs to imaginative educators that the new technology of pencils, paper and printing could have a beneficial effect on the schools of the country. Many suggestions are made. The most radical is to provide all teachers and children with pencils, paper and books and suspend regular classes for six months while everyone learns the new art of reading and writing. The more cautious plans propose starting slowly and seeing how "pencil-learning" works on a small scale before doing anything really drastic. In the end, Foobarian politics being what they are, a cautious plan is announced with radical fanfare: Within four years a pencil and a pad of paper will be placed in every single classroom of the country so that every child, rich or poor, will have access to the new knowledge technology. Meantime, the educational psychologists stand by to measure the impact of pencils on learning.

Seymour Papert, 1996, *Washington Post Education Review*

There is a growing trend for schools to equip every student with a computing device with wireless connection. The argument behind this generous investment is that for children truly to benefit from information and communication technologies (ICTs), each and every one of them needs to have a computer with him all all times. We do not expect a student to share a computer with others any more than we expect him to share his pencil. This argument is powerful and the image is appealing in both the developed and developing worlds. After all, we want our children to be able to learn at any time in any place, and we want them to have the best learning tools available when that "any time and any place" occurs.

Thus, schools began to put highly sophisticated digital pencils (computers) in the hands of children. Currently, at least 33 states[1] in the United States have one-to-one laptop programs, ranging from individual schools to school districts to statewide projects. Maine was one of the first states to start state-initiated one-to-one laptop programs. In January 2002, Maine signed a 4-year, $37.2 million contract with Apple Computer to install wireless Internet connections in all 239 public middle schools and to provide iBook laptops to every seventh- and eighth-grade student and teacher in the state—34,000 students and 3,000 teachers. Michigan followed suit. Michigan has spent about $30 million issuing a combination of laptop computers and personal digital assistants (PDAs) to nearly 30,000 students in 15 school districts. Started in 2004, Texas's Technology Immersion Project granted funds to 25 school districts around the state to provide not only one-to-one computing, but also "ALL of the tools they needed to conduct learning in the 21st century."[2] In January 2006, the governor of South Dakota proposed $13 million to provide laptops to every high school student.

Where states do not have the plan, individual schools or school districts have come up with the funds. In Ohio, for example, a middle school developed a one-to-one computing scheme with its own funding, which gave every child a computer. In Florida, the plan went further. A virtual high school was set up and the students of the school were given a computer to learn online without going to a school building. (Actually, there is no school building for this school.) In Virginia, the Henrico County public schools distributed over 25,000 wireless laptops to students and teachers spanning grades 6–12. Similarly, the Liverpool Central School District in New York, Community School District Six in New York City, and Clovis Unified School District in California also had one-to-one laptop programs.

The United States is, of course, not the only country that buys digital pencils for its children:

> In Ireland, a €2.7 million pilot initiative provided laptop computers to junior high school students with dyslexia or other reading and writing difficulties (Conway, 2005; Daly, 2006; Department of Education and Science, 2001).
>
> In Canada, the Enhanced Learning Strategy Project is putting laptops into the hands of all students from grades 3–11 in 26 schools and into an adult education center in Quebec.[3] The Wireless Writing Program (WWP) provided iBooks to all 1,150 grade 6 and 7 students in Peace River North (SD60) in 2003.[4]

In the United Kingdom, more than 250,000 laptops have been deployed to teachers nationwide under the Laptops for Teachers initiative (LfT).[5] Over 90 primary and secondary schools have Tablet personal computers (PCs), many on a one-to-one basis (Twining et al., 2006).

In New Zealand, a similar program has been implemented in Auckland schools.[6]

In Australia, the Victoria State Department of Education and Training earmarked Au$6 million (about U.S.$4.5 million) to supply each of the state's 42,000 teachers with a laptop. In Melbourne, all students from grade 5 and up in Wesley College have their own laptops with access to high-speed Internet and peripheral technologies such as printers, scanners, and CD burners.[7]

One-to-one computing has also spread to emerging economies and some developing countries. For example, in Singapore, the Crescent Girls' School gave 35 teachers and 355 students Tablet PCs on a one-to-one basis.[8] In Hong Kong, the Yau Ma Tei Catholic Elementary School equipped all of its fourth graders with a Fujitsu Tablet PC. Even in the poorest, remotest region of China's Yunnan Province, an elementary school started experimenting with this idea by requiring every student in one class to buy a laptop computer. The groundbreaking One Laptop per Child (OLPC) initiative is planning to provide U.S.$100 laptops for children in developing countries. The OLPC initiative was endorsed by U.N. Secretary General Kofi Annan at the World Summit on Information Society (WSIS) meeting in Tunisia in November 2005. The initial rollout of OLPC will focus on six developing countries (Young, 2005); five countries—Argentina, Brazil, Libya, Nigeria, and Thailand—had made tentative commitments by the end of 2006 (Markoff, 2006).

This list can go on and on. There are certainly other similar programs in place and one can reasonably expect that more are being planned. It is a good time to take a serious look at this phenomenon. We have enough case histories to illuminate the issues related to one-to-one computing, and what we learn can be of tremendous value to policy makers, educators, and members of the public who are either considering putting or already planning to put the digital pencil in the hands of many more children.

Therefore, this book takes a historical and international look at the digital pencil movement. Based on data drawn from a number of resources, we address the following questions:

Is the digital pencil a good idea? While the image of every child having his or her own computer is seductive and there are plenty of good arguments for doing this, the question of whether the computer can truly lead to better education is always at the backs of the minds of even the most enthusiastic supporters. After all, a computer is much more expensive than a pencil or book, and evidence showing large-scale gains of using technology in schools is still lacking. At a time when school funding is decreasing, the money can definitely be used for something else—raising teacher salaries, reducing class size, or repairing the roof, for example. Unless the computer can result in dramatic gains, it is worth reconsidering the value of one-to-one computing. Furthermore, stories about children addicted to surfing the Internet, playing computer games, and chatting with strangers and friends online pose problems and reveal the potential harms of one-to-one computing. Additionally, reports about the underutilization of existing technology available to teachers and students add to the mounting evidence against putting more gadgets in schools. Thus, a critical examination of the costs and benefits of one-to-one computing becomes necessary.

The cost-effectiveness analysis is, however, not as straightforward as we would like because the cost is not only the expense of bringing the computer to the child but also the costs for making it usable, including connectivity, electricity, technical support, maintenance, software, and constant upgrades. There are also other costs: teacher training, student time, potential psychological and developmental harms, opportunity cost to the school, and opportunity cost in terms of child and adolescent development. Likewise, the benefits are not easy to calculate. They can be short term: student engagement in learning, student technology proficiency, access to newer materials, and the capacity to communicate with others. They may also take a while to realize: more positive attitudes towards schooling, better academic achievements, and, perhaps, an overall better person. The benefits can be functional and utilitarian: direct impact on teaching and learning; they can also be purely symbolic: making the school appear progressive and on the cutting edge.

In this book, we analyze the costs and benefits of one-to-one computing programs by taking into consideration as many indicators as possible. We will also examine the evaluation reports of various projects within our analytical framework to present a comprehensive summary of outcomes of one-to-one computing projects.

What happens when each child has a networked computer? Policy makers, teachers, parents, and the public are curious about the changes one-to-one computing will bring to the lives of children. What do they do with it? We know that technology can be used for good and for bad. But what are "good" uses and what are "bad" uses? Do students use the laptops for good or for bad? When and under what conditions are good and bad uses more likely to occur? What do teachers, parents, and schools need to do to encourage good uses and discourage bad ones? What content is necessary? What kind of teacher training is needed? We will analyze existing data to answer these questions in the hope that we can gain some insight into these questions and come up with suggestions and recommendations for policy makers, teachers, and parents when they plan to buy each child a computer.

What is a "better" device? Diversity and rapidity of change are trademarks of technology. Computing devices can range from a $50 second-hand Palm Pilot or other similar handheld devices, often called PDAs, to a $2,000 laptop computer. Of course, they differ in what they can do as well. There are plenty of arguments over what is an ideal device for students: PDAs are inexpensive, easy to carry, and powerful enough for most learning tasks, but their screens are small, their keyboards are not easy to use, and they cannot perform high-end tasks such as video editing and transmitting. A laptop computer, on the other hand, is more versatile, has large screens and keyboards, and has much more processing power, but it is expensive and more prone to problems. What should schools buy?

Technology changes very rapidly. A computer becomes outdated very quickly. Therefore, how to keep the plan going is another burning question. Policy makers often ask whether schools should lease or buy the device. Leasing provides more flexibility and seems to cost less annually, but buying a computer frees the school from further financial commitment for at least a few years. In this book, we examine the relative advantages and disadvantages of different devices and implementation schemes.

How do we know if this is making a difference? Because schools are accountable to the people who spent the money buying the computers, evaluation becomes necessary. But what kind of evaluation makes sense? As mentioned before, the benefits and costs of technology in schools are often not obvious; thus, we will need an evaluation framework that can comprehensively and systematically capture the efforts and assess their impact. In addition, given the pioneering nature of these programs, we should learn from them so as to inform others and ourselves in the future. Hence, a serious research component should be part of the evaluation. In this book, we will review the evaluation plans of the various projects and propose a framework for comprehensive evaluation and research on one-to-one computing.

What is the future of one-to-one computing? The penetration of information and communication technologies into classrooms has created opportunities for ubiquitous computing to take root in schools and affect the established cultural practices. Schools in turn shape what ubiquitous technologies are used, in what ways, and for what purposes. The future of one-to-one computing is currently emerging, pushed by three major forces: the penetration of technology in our lives, the increasing demand of using mobile technology anytime anywhere from the digital generation, and the continuous investment in school technology with rising expectations about education's role in fostering the knowledge society. These three forces, working from bottom-up and top-down simultaneously, are pushing schools from N:1 computing to 1:1 computing to 1:N seamlessly ubiquitous computing. In this book, we summarize emerging technology trends and define the role of one-to-one computing in preparing students with digital citizenship for a globalized education.

The primary audience of the book includes school administrators, educational technology professionals, researchers, and policy makers in the United States and elsewhere. The book can also be included as supplemental reading for advanced level graduate courses in education, technology, and technological innovation.

Data Source

The data in this book, both theoretical and empirical, were drawn from a variety of sources, including the research literature, existing reports, and a number of specific projects. The book's authors have each led a longitudinal study of a laptop project (one school level, one state level, and one national level). These projects were funded differently and implemented differently. Data were also drawn from other ubiquitous computing projects in multiple countries through the authors' personal contacts and review of the literature. The projects include the Wireless Writing Program (WWP) in the Peace River North school district in Canada, the Tablet PCs used in U.K. schools, the Cyber Art Project in Hong Kong, the pocket PCs with wireless networks program in Chile, the Laptops Initiative for Students with Dyslexia and other reading and writing difficulties in Ireland, and several ubiquitous computing projects in the United States. The U.S. projects were conducted at three levels: statewide projects such as the Michigan Freedom to Learn (FTL) project, Maine Learning Technology Initiative (MLTI), and the New Hampshire Technology Promoting Student Excellence (TPSE) project; school-district-level projects such as the Henrico County Public Schools (HCPS) in Virginia and the Quaker Valley Digital School District in Pennsylvania; and school-level projects such as Alpha Middle School project and Rye Country Day School project. A detailed description of these projects can be found in appendix A of this book.

Theoretical Framework

In this book we use the ecological perspective as an overarching framework to guide our discussions. An ecological perspective for examining technology use in schools has been well articulated by Zhao and Frank (2003). They maintain that a school and its classrooms can be viewed as an ecosystem within which the characteristics and roles of different species continuously affect one another and constantly change their relationships. They establish metaphorical equivalents between technology uses in schools and ecological issues:

Schools as "ecosystems": A school ecosystem is a combination of diverse parts and various relationships. It consists of abiotic components such as school buildings and classroom equipment, and biotic components such as students, teachers, administrative staff, technology staff, and a school board. These species are closely connected to each other and the relationships are very complex. In addition, a school system has many different resources and materials that allow for individual species' proclivities and interests.

Computer uses as "living species": Technology use can be viewed as a living species because, like any other species, it has a niche: the role it plays in a school ecosystem. It needs resources such as money, technology support, and training in order to grow. It interacts with other species; it influences how teachers teach and how students learn and also affects school social relationships. It evolves through interactions with other species and the environment.

Teachers and students as members of a "keystone species": Teachers and students are the most important and essential components of a school system. They play a crucial role in the process of using technology in teaching and learning because they determine whether, what, and how technologies are used.

External educational innovations as invasions of "exotic species": They break the equilibrium of the ecosystem, compete with other species, co-adapt, and co-evolve.

We believe that the ecological perspective is a powerful analytical framework to capture the dynamic nature of technology use in schools holistically. The technologies we discuss in this book are so enmeshed in daily activities that they "disappear" (Bruce & Hogan, 1998). Thus, only an ecological perspective can give us a basis for understanding the interpenetration of and mutual influence among technologies, human beings, and the context within which they interact.

Chapter Overviews

This book starts with a historical view of information communication technology for children. In the wake of the apparently inexorable drive to adopt the latest shiny device—especially mobile technology—chapter 1 adopts historical and ecological perspectives in order to examine the progress in educational technology inventions and critically read the discourses and policies promoting educational technologies and evaluations of the diffusion of technology in educational settings. Today, the shift from an innovation to an appliance view of technology accurately characterizes changing views of technology in schools and society. We use this ecological view to examine the implications of this shift for ownership (by groups and individuals), shrinkage and growth of devices, diffusion in educational settings, and the functions of social capital in technology diffusion.

Is ubiquitous computing a good idea? From different perspectives, different stakeholders have made various arguments for the benefits it can bring. These arguments have their own strengths and limitations. Chapter 2 introduces these arguments and examines their plausibility. In describing and appraising the various arguments used to justify significant investment in one-to-one computing, we want to highlight the powerful symbolic role played by arguments for technological innovation in education. In addition, to better evaluate the rationale for one-to-one computing, this chapter discusses its costs at different levels, including the costs of bringing the computer to the children, expenses for making it usable, and other opportunity costs.

To make this idea work, various issues need to be taken into consideration. Chapter 3 examines issues that affect how laptop use shapes and is shaped by the school culture in relation to technology use by students, teachers, and principals. We examine laptop deployment and management models within specific school ecologies and how these ecologies evolve over time. In particular, we focus on the manner in which locally defined educational values and priorities can change the intended uses of laptops. We highlight the role of school culture in shaping the innovation–appliance relationship by focusing on existing school practices in relation to technology and innovation, school leadership, and wider educational system learning aims and goals.

What happens when each student has his or her own laptop? In chapter 4 we examine the ways students use laptop computers in various contexts and for various purposes—some for learning, some for entertainment or social reasons, some expected and supported by adults, some discouraged by adults but loved by students, some with positive consequences, some with negative consequences, and some with complicated, not so clear-cut consequences. Through examining student laptop uses from different angles and at different levels, this chapter discusses the many different roles the laptop computer plays in students' lives.

Among various one-to-one computing devices, what is a better choice and in what context? To answer these questions, chapter 5 compares different one-to-one computing devices, especially laptops with handheld devices, and examines a number of considerations that shape debates in schools about the merits of particular laptop or PDA purchases, including cost, symbolic appeal, management, learning, and support. We note the way in which these considerations echo dynamics of introducing other types of innovations in education.

Students' having their own laptops not only influences how they learn and interact in schools, but also changes the dynamics at home, which is a critical part of a student's development and exerts further impact on student learning in schools. To understand the penetration of one-to-one computing in student lives fully, chapter 6 examines the mutual influences between one-to-one computing and student activities at home, their relationships and interactions with parents and siblings, and other aspects of the home environment.

Focusing on the evaluation of diverse laptop initiatives, chapter 7 examines how we know if one-to-one computing is making a difference. Given the significant investment and their innovative nature, most laptop initiatives have some sort of evaluation plan. This chapter reviews the various evaluation approaches and findings of different laptop projects that have been undertaken in a variety of contexts, seeking to understand the impact of laptop initiatives with varying aims. In addressing the diversity of laptop projects, case studies focus on local-school, special-needs, system-wide, home-school, and after-school contexts. Chapter 7 asks what we can learn from specific evaluations as well as how we might frame evaluations in order to ask the right questions to envision the design of future evaluations. As evaluations of one-to-one computing proliferate, we think it is important for them to address not only the specific goals of particular initiatives but also the wider context of educational reform, which is increasingly a feature of efforts to reshape education systems to support the knowledge society (Hargreaves, 2003).

Few people would challenge the claim that technology is inexorably creeping into our lives. This book focuses on this phenomenon by examining the discourses, initiatives, and impact of the digital pencil phenomenon in the lives of K–12 students in the context of children's and adolescents' media consumption habits. At its most basic, the book examines laptops in education: their justification, use, and impact on learning, learners, and identity. As such, it focuses on a key technological innovation and its role in educational change. However, this book also can be read as a case study of the human–machine relationship in learning contexts. It is within this wider perspective that we offer a radical and powerful vision of the future of educational computing: educational technology co-evolution and co-adaptation for digital citizenship. This vision situates the digital pencil phenomenon in the context of reshaping the school–society relationship in order to prepare students for digital citizenship in an era of globalization.

Acknowledgements

We would not have been able to write this book without the support of our colleagues, families, friends, students and research collaborators in China, Ireland and the USA. In particular, we want to thank the many educational policy makers, school leaders, teachers, and K-12 students with whom we have worked in various different settings. Their insights have helped us understand the dynamics of one-to-one computing in diverse educational settings.

We thank our colleagues at Syracuse University, University College, Cork (UCC) and Michigan State University who provided support for this cross-national collaboration. In particular, we are grateful to the College of Celtic Studies, Arts and Social Sciences at UCC which provided funding for editing and indexing. We especially want to thank Simon Coury whose careful proofreading and indexing made the task of writing this book a lot easier.

We are especially grateful to Naomi Silverman of Lawrence Erlbaum Associates and Judith Simon of Taylor & Francis for their support.

About the Authors

Dr. Jing Lei is an assistant professor in the School of Education at Syracuse University. She received her PhD in learning, technology & culture from Michigan State University in 2005. Her research interests include educational technology integration, meaningful technology use in schools, social-cultural and psychological impact of technology, teacher educational technology professional development, and international and comparative education.

Dr. Paul Conway is a college lecturer in the Education Department at University College Cork, National University of Ireland (www.ucc.ie) and visiting scholar at Michigan State University. Prior to that, he was an assistant professor of educational psychology and human development at Cleveland State University. He is a graduate of St. Patrick's College, Dublin (BEd), Boston College (MEd) and Michigan State University (PhD). He has extensive experience in the areas of learning, literacy, and ICT/educational technology policy.

Dr. Yong Zhao is university distinguished professor in the Department of Counseling, Educational Psychology, and Special Education at the College of Education, Michigan State University. He serves as the executive director of Confucius Institute and is the founding director of the Center for Teaching and Technology and the U.S.-China Center for Research on Educational Excellence. Dr. Zhao received his PhD in education from the University of Illinois at Urbana-Champaign in 1996. His research interests include diffusion of innovation, teacher adoption of technology, computer-assisted language learning, second language teaching, and globalization and education.

A historical view of technology in schools 1

In the nineteenth century, there were no televisions, airplanes, computers, or space craft; nor were there antibiotics, credit cards, microwave ovens, compact discs, or mobile phones.

There was, however, an Internet.

Tom Standage, *The Victorian Internet*, 1998

We should not be surprised, then, to find in the history of information technology, and in its current configurations and future projections as well, an evolutionary dynamic in many respects much like that of the literally natural, organic world.

Paul Levinson, *The Soft Edge: A Natural History and Future of the Information Revolution*, 1997

The most profound technologies are those that disappear. They weave themselves into the fabric of everyday life until they are indistinguishable from it.

Mark Weiser, *Xerox computer scientist, 1991*

Introduction

In the history of human development, numerous technologies have been invented and put into use; many of them have been introduced into classrooms since the first schools were built, from chalk and blackboard in the early days to film, TV, and the computers of recent years. While some of them survived the competition and selection, most technologies disappeared soon after their invention or introduction. Is ubiquitous computing just another fad that will soon fade away, or is it here to stay and therefore worth investing in? To better understand the nature of technology adoption in schools, we need to examine contemporary technology integration from a historical view. In this chapter, we first review the evolution of technology from innovations to appliances and the implication for school technology adoption. We then discuss the major trends of the development of information and communication technologies, examine the adoption of technology in schools, and finally return to the questions of whether or not ubiquitous computing in schools is a good idea and what schools can do with it.

Innovation and appliance: the evolution of technology

In his book, *Diffusion of Innovations*, Everett Rogers (1995) describes the process of innovation reaching different groups of people, from a small number of "innovators" to "early adopters" to "early majority" and then finally to "late majority." In general, technologies that reached the early majority and late majority stages have gone through a transformation from innovations to appliances (Zhao, Lei, & Frank, 2006). An innovation is something new or the act of "introducing something new."[1] Innovations are normally rare, expensive, unstable, and unreliable and they have functions that are often uncertain and evolving. Compared to innovations, appliances are more affordable, widespread, and reliable; they have fixed functions and often disappear into the context where they are used.

Many of the technologies we use every day and take for granted were rare innovations when they first appeared and have gone through this evolutionary transformation. The history of the automobile is a good example. When the first cars were made and put on the roads, they were rare innovations that needed special treatment. For example, the Red Flag Act of 1865 in Britain required a person to precede a car on foot, waving a red flag to warn horse riders and pedestrians of the vehicle; it set the speed limit at 4 mph (2 mph in towns).[2] Nowadays, cars have become a natural and indispensable part of many people's lives, especially those in developed countries. Compared with the special treatment in the early days, we do not even mention cars when we talk about the daily activities that involve them (e.g., "I'll take you to the airport") or we use driving time to talk about distance (e.g., "Washington, DC, is 4 hours from New York City").

Once a prominent and rare innovation, cars have seamlessly disappeared into our daily lives.

The shift from an innovation to an appliance view of technology accurately characterizes the changing role of technology in schools and society. Here, we use this perspective to examine the implications of this shift for ownership (by groups and individuals), shrinkage and growth of devices, distribution of expertise and social capital, diffusion in educational settings, and impacts on schools.

From group ownership to individual ownership

The transformation of technology from innovations to appliances is generally accompanied by an ownership transfer from group ownership to individual ownership. Technologies can belong to groups and to individuals. Some technologies are generally owned by groups, such as space shuttles, commercial jets, and large infrastructures. Most people cannot afford and do not need to own an airplane or a commercial train. Some innovations are more likely to be owned by individuals, such as pencils, personal computers, and cell phones. Most technologies experience an ownership transfer from groups to individuals as they transform from innovations to appliances.

Group ownership is much more common than individual ownership at the early stage of an innovation, generally due to the high cost of ownership, paucity of resources, or lack of expertise. For example, a few decades ago, computers were very expensive to operate and mostly belonged to companies or universities. The IBM 701 EDPM in the 1950s cost $15,000 per month to rent and all 19 of them went to research institutions, companies, and government agencies.[3] During the early days of telephones, it was common for a whole company or an entire building to own one telephone. A specific person was assigned to a telephone machine—answering phone calls, taking messages, fetching the person for whom the phone calls were intended, helping to place a phone call, and, in some cases, collecting a fee for the phone call and the service. People sometimes went to the post office to make a phone call and professional assistance was provided by an operator.

Gradually, phones became more available. A few years ago, phone booths were everywhere in the streets of many big cities worldwide, from Shanghai to New York City, and prepaid phone cards were available for sale at every street corner. Today, mobile phones have become commonplace. Nearly 70% of adults and half of teenagers in the United States had their own mobile phones in 2005 (Lenhart, Madden, & Hitlin, 2005, p. 10). In China, there were approximately 459 million mobile phone users by the end of 2006 (*Xinhua English,* 2006). It is common for people to have an office phone, a home phone, and a mobile phone; some people even have more than one of each.

Cars have also experienced dramatic ownership change. Once expensive luxuries for a few wealthy people, cars are now owned by almost every family in developed countries. The National Transportation Statistics 2005 reported that the United States had about 243 million registered motor vehicles in 2004,[4] which means that, on average, almost every person owns a motor vehicle.

Individual ownership provides much more convenience to users. When telephones were owned by organizations, people wanting to use a telephone had to coordinate with each other and probably wait in line. Making a phone call today is as easy as taking a breath. You can hear all kinds of mobile phones being turned on immediately after a plane has landed and see people glued to their phones while walking down the streets or waiting in line to get a cup of coffee; very likely you are one of them. Similarly, cars provide people with the freedom to go wherever they want at their convenience.

However, there are trade-offs. Individual ownership costs much more and consumes more resources. For example, the number of motor vehicles in the United States tripled between 1960 and 2004, as did fuel consumption (National Transportation Statistics, 2005). Mass individual ownership of particular technologies can even cause inconvenience to the very services they aim to provide. For example, in most of the world's urban areas, individual car ownership has resulted in traffic jams and pollution. The negative effects of individual ownership have led to a return to group ownership. For example, many cities now have incentives for car pools or using public transportation. Other cities try to restrict existing individual ownership; many large cities in the world have policies and regulations to discourage individual car ownership. In London and Singapore, for example, drivers pay a fee to enter the city.

Individual ownership of some technologies may also create much more waste than group ownership does, thus posing serious challenges to the environment. It is estimated that up to 50 million metric tons of "e-waste" are generated worldwide every year (Blau, 2006). In 2005 about 133,000 personal computers (PCs) were retired every day in the United States alone (*U.S. News & World Report,* 2005). Given the increasingly rapid development of technology, it is reasonable to assume that this number is growing. Millions of outdated or broken computers, monitors, televisions, and mobile phones ending up in landfills are a serious threat to the environment because the electronic waste may contain hazardous materials.[5]

To address the environmental hazards of waste disposal—specifically, management and disposal of old computers, many countries and regions, including South Korea, the European Union, China, Japan, the United States, and Taiwan, have established specific regulations concerning e-waste (*Computer Economics*, 2006). In November 2006, delegates from about 120 governments attended a United Nations summit to discuss reducing the use of toxic substances and promoting programs for recycling and reusing as many components as possible (Blau, 2006).

From big to small and from small to big

One reason that many technologies can expand their ownership from groups to individuals is that technology innovations change dramatically in their size while meeting and expanding people's needs. Many technological devices have dramatically shrunk in size during their development. For example, motors have disappeared inside many small electronic devices (Dede, 1996). Mobile phones also shrank from big "bricks" in their early days to small, thin devices that can fit into a pocket. However, probably none of these size changes can be compared with the remarkable change in size of computers. The first computers were enormous pieces of equipment filling entire rooms. The famous Harvard Mark I in 1944 was a 5-ton gigantic device, about 55 ft long and 8 ft high, that contained almost 760,000 separate pieces and about 500 miles of wire and "sounded like a long room full of old ladies knitting away with steel needles."[6]

The "room computers" gradually shrank into desktop computers. The first Apple Macintosh in 1984 weighed 16.5 lb, which was hardly "portable" but much more convenient to use and maintain, and it was getting closer to laptops. Twenty years later, laptops generally weigh a few pounds and can easily fit into a briefcase. Small laptops such as the Sony VAIO of 2004 are less than 2 lb. Handheld computing devices are lighter than an average book. A Compaq handheld PC is only about 5 oz. People put their laptops into their backpacks and carry their personal digital assistants (PDAs) in their pockets. Portability has made laptops and handheld devices increasingly popular. In May 2005, 53%

of computers sold in the United States were laptops, surpassing desk-tops in market share (*Newsweek*, 2005).

As computers shrink in size, their displays are getting larger. Com-puter displays come in different sizes, but a general trend is for them to be increasingly bigger relative to the processor. Now the small moni-tors for laptops can be as small as 7 in.; some small desktop monitors are 14 in. or smaller. More common ones are 15, 17, or 21 in. and large monitors can be 45 in. or bigger. Another way to have a bigger monitor is to have multiple monitors simultaneously displaying multiple pages, windows, applications, graphics, full-screen video, and games.

The variety of sizes provides technology users many more options that, on the one hand, meet users' various needs. If you are a light traveler, you probably want a small laptop with a 12-in. or even smaller screen. However, if you want to display a graph on your computer to a group of people, you certainly want a monitor as large as possible. On the other hand, more options may also increase the need for using technology. For instance, only when portable computers were available did people feel the need to take their laptops with them wherever they traveled, even on vacations. Users' various needs increased the diver-sity of technology, which in turn changed users' needs. In this sense, users and technology co-adapted to each other and co-evolved.

From simple to complex and from single to multiple functions

Technologies change not only in size, shape, and weight but also, more importantly, in their functions, which are becoming increasingly com-prehensive and sophisticated. For example, the LeapPad learning sys-tem developed from a simple electronic board to a talking book with voices, music, sounds, and games—from a simple educational toy to an interactive learning system on which children can perform differ-ent activities and play songs and games. Now LeapPads can also be connected with computers and used to download information from the Internet. Similarly, a PSP (play station portable) can have built-in Wi-Fi and can play videos and songs. It can also display photos and home movies, stream television shows from a home network, and even be used as a voice-over-Internet phone.

The mobile phone is another telling example. Mobile phones were just a mobile telephone device a few years ago. The only function was to communicate messages through voice. Now mobile phones have multiple functions and are becoming increasingly complex. In addition to the original calling function, a mobile phone can be used for searching and downloading information online; receiving, typing, and sending e-mails; sending text messages; taking pictures; recording videos; playing games; scheduling events; computing; programming; and much more. The mobile phone has become so sophisticated that it is replacing other technologies, such as the address book, planner, camera, tape recorder, and voice mail; thus, it has become indispensable for many. In a poll conducted by Lemelson–MIT Invention Index in 2004, the mobile phone was voted the invention that people hated the most but could not live without.

The functions computers can perform have also been multiplying. Once a device mainly for mathematical calculations, the computer today remains a computer (it still does the computing), but it is also a camera (actually, a photo and video studio complete with film-development tools and chemicals), television, DVD player, radio, newsstand, piano, game board with many games, phone, fax machine, travel agent, and much more.

Specific computer programs have added more functions as well. For example, a word-processing program used to be just an electronic typewriter. Now it has more than 1,000 different specific functions, including composing, editing, and even multimedia producing, just to name a few of the most frequently used functions; the list could go on and on.

From unreliable to stable

When a technology is still at its innovation stage, it is often unstable, unreliable, and easy to break. Not many people understand it or know how to use it, so it is difficult to get support for it. The inventors and early adopters are still trying to figure out what to do with it. Revisions and changes are often made, so new functions are always evolving. Gradually, the functions are stabilized and improved, more people start using the technology, and more people know how to use it. The more widely a technology is diffused, the more widely the expertise about it is distributed.

More widely distributed expertise makes it easier to obtain help and thus increases social capital, which further facilitates the diffusion of a technology innovation. Social capital is the potential to access resources through social relations (Coleman, 1988; Putnam, 1993; also see Lin, 2001; Portes, 1998; and Woolcock, 1998, for recent reviews). Frank, Zhao, and Borman (2004) argue that a person who receives help that is not formally mandated draws on social capital by obtaining information or resources through social obligation or affinity.

Social capital comes from people's social network, which plays a vital role in their technology adoption. As social beings, people are easily influenced by others around them. If a person is highly interconnected in the social networks, he or she is more likely to be exposed to innovations in technology and more likely to be influenced by innovative people (Rogers, 1995, pp. 273–274) who have expertise in using certain innovations and thus can provide help and support to people in this social network. A social network also works through peer influence. On the one hand, peers who adopt a technology innovation can provide expertise and technological and human support, while at the same time helping to create an encouraging environment for the adoption of innovations. On the other hand, peers' technology use also makes one feel the urgency to catch up in order to keep one's sense of competency and of belonging because peers exert psychological pressure for group conformity (Frank et al., 2004)

As a technology innovation becomes more accessible to the general public, expertise distributes to a wider population and the users' social capital grows richer. People can learn from each other and assist each other; this pushes the further adoption of the technology, changing it from an innovation to an appliance.

ICT trend

The evolution of information and communication technology (ICT) follows the pattern of technological evolution: change in ownership, size, function, expertise, and social capital. It also has some prominent development trends: It is becoming ubiquitous, the functions are converging, and it is increasingly interactive.

Over a billion internet users: the ubiquity of ICT

The transformation from innovations to appliances can be very rapid or very slow, depending on the specific technologies and how they interact with the users and social contexts. Some innovations can take centuries to popularize; some can finish the transformation overnight when they reach a "tipping point" (Gladwell, 2000).

The transformation for ICT is a dramatic one because it has experienced tremendous development in a very short period of time. Only a few decades ago, computers were very expensive to operate and maintain; they were owned by companies or universities and only accessible to a few experts. Now they have become increasingly portable, powerful, efficient, reliable, and ubiquitous. We use computers at work to write reports, teleconference business partners, complete financial transactions, and get information. We also use computers at home to find directions to restaurants, locate recipes, play games, call friends, and shop for everything from books, movies, and songs to clothes, TVs, and cars. In 2002, 70% of American travelers did travel research online, and more than half of them booked reservations on the Web, spending $22.6 billion (*Newsweek*, 2002). In 2004, there were more than 800 million Internet users around the world,[7] and this number increased to about 1,086 million as of September 2006.[8] Among them, more than 209 million were American Internet users,[9] who accounted for 73% of the U.S. population (Madden, 2006), spending an average of 3 hours a day online (Stone, 2005) and approximately $170 billion in 2006.[10]

ICT has also moved into school buildings and classrooms. Today, schools in the United States have the most technology access in the world. Computing technology has become part of many children's daily lives. According to a report released by the National Center for Education Statistics in 2003, overall, 91% of students from nursery schools to high schools used computers and 59% used the Internet. The use of these technologies began at young ages and increased as the grade level advanced: 67% of children in nursery school were computer users and 23% of them used the Internet; in middle schools, 95% of students used computers and 70% of them were Internet users.

By high school, nearly all students (97%) used computers and a majority (80%) used the Internet (NCES, 2005); this percentage rose to nearly 100% in 2004 (Levy, 2004). The scope of teens' online lives is broad, from searching information for schoolwork to chatting with friends, from playing online games to making online purchases, and from accessing news to seeking health information. In a poll conducted in 2002, among the six most popular media, Internet was chosen by one third (33%) of children aged 8–17 as the medium they would want to have if they could not have any others, followed by TV (26%) and telephone (21%) (Knowledge Networks/Statistical Research, 2002).

These figures suggest that, over the last decade or so, computing technology has evolved from an innovation reserved for a few to an appliance accessible to and used by many. This evolution changed not only the meaning and characteristics of computer technology and the role it plays in our lives, but also how we should view technology and education. As innovations, computers were rare and unreliable. Because of their novelty and rarity, we were mostly interested in trying to explore what, if anything, computers could do to help education; thus, the central question at the early stage was whether computers improve educational outcomes.

To answer this question, research focused on designing and developing specific uses of computers to affect student learning. The view of technology and education at this stage was primarily psychological; its focus was on the psychological effects of computers on the learners. When a particular software or use of computers was found or thought to be effective in one situation, efforts were then made to extend the benefits to more people. At this stage, views on technology and education took on a sociological perspective and began to be concerned with how best to duplicate success stories in more classrooms. Hence, for a while, a popular move in many nations was to prepare teachers to be technologically proficient so as to implement "best practices" invented by researchers or early adopters in their own classrooms.

Nowadays, computer technology is no longer an innovation. Computers have become appliances seamlessly integrated into our daily lives. Therefore, the view of information technology is shifting towards an ecological one, taking it as an integral component of the context where it disappears.

Web on the TV and TV on the Web: the convergence of functions

Ofcom, a U.K.-based communications agency, conducted a worldwide study on the world communications market. The 2006 report points out the convergence of devices as one of the three worldwide convergence trends (the other two are billing convergence and platform convergence). Device convergence "allows consumers to access many different services from the same device, even if they are delivered over different platforms" (Ofcom, p. 14, 2006). Many technologies originally had one or two functions and served one purpose. Television sets showed TV programs, copy machines made copies, cameras took pictures, DVD players played movies, and blackboards displayed written information. In the digital era, many functions can mix and converge in one device. In addition to making copies, a copy machine can fax, scan, print documents, and sort and staple the documents when needed.

Thus, it is often difficult to say what a device can do. For example, now TV watchers can watch their favorite news channel or sitcoms even if they do not have a TV set, as long as they have a computer and Internet access, because many TV programs are available on the Internet. Some Web sites offer ad-supported downloads of old TV shows, and some TV networks such as NBC and CBS make some current shows available on demand for a fee.[11] On the NASA TV Web site, TV programs include coverage of live events, news, special programs, regular public and media programs, the NASA television education channel, and the NASA television video file. Viewers can watch real-time TV programs or download program clips. In fact, according to a worldwide study by Ofcom, broadband users are more likely to watch TV programs online. On average, one third of broadband users watch less television since going online, with Chinese broadband users leading the trend: 70% of them watch TV over broadband (Reuters, 2006)

Meanwhile, Internet users do not have to have a computer to surf online. Now many computer technology applications have been added to TV. A cable TV can do at least the following things:

- display TV programs;
- play back DVD-quality video;
- play CD-quality audio;
- stream a personal library of PC-stored MP3, WMA, and WAV audio files;
- play Internet radio and commercial music services on the stereo;
- connect to home computers and display digital photos;
- display real-time audio and video at teleconferences;
- surf on the Web;
- chat online;
- send and receive instant messages; and
- play video games.

At the same time, all these things can be accomplished on a computer. Connected to the Internet, computers are becoming increasingly multifunctional. As Dede (1996) points out, in the past there were many different technologies (similar to individual species) crowded in an organization, known as "information ecologies" (Nardi & O'Day,

1999). Now, different technologies are fusing together; "the radio, television, telephone, copier, fax, scanner, printer, and computer will eventually coexist in a single box (Nardi & O'Day, 1999)." Therefore, according to Dede (1996, online), it is possible that the ecology of information technologies will gradually have only a few "superspecies" that synthesize and extend the capabilities of all current devices.

From push to pull: on-demand technologies

Another ICT trend is the shift from "server push" technology to "client pull" technology. In the past, access to information via radio and television was driven by broadcaster schedules, and users did not have much control over what information they could obtain. Now, users can increasingly choose when, where, and how they may "pull' from a vast array of streamed and archived digital information. For example, Korean teenagers are leading the world in their mobile-phone viewing of on-demand TV programs. The capacity to use the Internet as a TV program archive (especially for news and current affairs programs) reflects a similar trend on the Internet that, with the booming of Web 2.0 and search engines, is becoming increasingly interactive and "on demand." This shift from "push" to "pull" encourages more active involvement of users and allows for more interactivity between users and technology.

From reader to writer: personalized publishing and broadcasting

The most fundamental change in ICT is perhaps the capability it affords individuals to publish and broadcast their ideas to a broader audience. Thanks to low-cost digital tools and easy access to the Internet, practically anyone who wishes to publicize his or her ideas, images, or any other personal information can do so. Publishing and broadcasting are no longer controlled by corporations or organizations; they are now within the reach of the individual. Podcasting, Web logs (or blogs), and YouTube.com are just a few examples of personalized publishing and broadcasting.

Podcasting, simply put, is broadcasting on the Internet. The technology is widely available and easy to use, making it possible for individuals to operate their own radio and TV stations. Podcasting was just a concept in 2000; the enabling technology became available in 2001. In 2006, the number of podcasts reached 44,000, surpassing the number of radio stations in the world.[12] The editors of the *New Oxford American Dictionary* declared "podcasting" the 2005 word of the year.

Blogging is another form of personalized publishing that has grown exponentially in a matter of years. A blog, or Web log, is a user-generated Web site written in a journal style. It started as manually entered online journal entries in the mid 1990s, but toward the end of the 1990s, tools to facilitate the creation of blogs became available, which made it possible for anyone who can browse the Web to contribute to the Web. Blogs began to grow at a quickening speed. As of November 2006, a conservative estimate put the number of blogs at 60 million. Blogs kept by individuals have already become an important competitor to traditional mass media such as newspapers.

Diffusion of technology in schools

From when chalk and blackboard entered schools centuries ago to today's "superspecies," the diffusion of technology in schools has gone through a tortuous course. Researchers and educators have long tried to incorporate technology innovations into school settings and help students take advantage of the newest technology, which, in the last few decades, has been mostly related to information and communication technology.

Computers entered schools starting in the early 1960s, and researchers with access to the mainframe computers of the time began experimenting with the possibilities of using computers for learning (Collis et al., 1996, p. 4). The technology brought into schools at that time was primitive, and computers in schools were accessible to only a small number of people. In the late 1980s and early 1990s, with the emergence of PCs and user-friendly graphic user interface (GUI), computers became more accessible to the general public and to schools. Many schools built computer labs and started equipping classrooms with computers.

The 1990s witnessed a rapid rise in investment in and advocacy of technology in schools. Several countries published their first national educational technology plans and standards in the 1990s and many developed countries currently have two or more national technology plans (Zhao, Lei, & Conway, 2006). In 1996 and 2000, the U.S.

Department of Education published its first and second national technology plans. Both contained specific goals to increase technology access in schools. The investment that many countries, especially the developed countries, put into schools to equip classrooms with technology has been staggering. For example, in Ireland, the second national ICT plan in 2001 invested IR £107.92 million in putting ICT into schools (Zhao et al., 2006). Over the last decade, the United States has spent at least U.S.$66 billion on technology in schools (Quality Education Data, 2004).

As a result of generous investments, both the quantity and quality of technology access in public schools have increased dramatically. Schools have become high tech, and teachers and students have increased access to advanced technologies. In the past two decades, the national student–computer ratio in the United States has dropped from 125:1 in 1984 to just 3.8:1 in 2004 (Technology Counts, 2005). Many schools have launched one-to-one computer projects and more schools are following. Internet access has also increased dramatically. In 1994, only 35% of public elementary and secondary schools and 3% of public school classrooms had access to the Internet. By 2004, nearly all schools had Internet access and 99% of instructional rooms had access to the Internet (Technology Counts, 2005). As the amount of technology accessible to teachers and students continues to grow, the variety of technologies available in schools has increased as well. Students now have desktops, laptops, handheld computers, peripheral technologies, Internet resources, multimedia technologies, e-learning systems, and numerous types of software that can be used for every subject area.

This transformation of ICT from innovations to appliances changed not only the meaning and characteristics of computer technology, but also the role computers play in schools. In the first decade or two, computers were mostly used for research purposes or administration. But educators quickly developed broader applications for computers in schools to facilitate teaching and learning. The title of Robert Taylor's 1980 book, *The Computer in the School: Tutor, Tutee, Tool*, reflects the use of computers during that period of time. The term "tutor" indicates that computers were used for drill and practice, tutorials, and simulations—the most common image of the computer in education at that time (Collis et al., 1996, p. 4). As a tool, computer software extended users' ability through word processing, database, painting and drawing programs, and Internet browsers; a computer could also serve as a tutee, if one thinks about programming as teaching the computer.

In the 1970s and early 1980s, technology was viewed mostly as a subject: Students were expected to study the machine and learn specific skills to make the technology work, and researchers studied the social implications of technology use in classrooms. Programming was taught in schools and some advocates of teaching programming argued that the important component of programming was to help students express an understanding of a problem or an idea by means of a simple program (Luehrmann, 2002). Papert's *Mindstorm* (1980) and the Logo programming language facilitated a grassroots movement that generated many thousands of classroom implementations of progressive education technology (Papert, 1992, p. 42).

When computers became more accessible with the advent of personal computers and GUI, they were viewed as a supplement to teaching and learning; curriculum-based software was developed and introduced into schools, and computers were considered as tools for the teacher and student. In the late 1990s, the rapid development of the Internet stimulated another dramatic growth of interest and a tremendous shift of content in computer education. Schools began to teach networking, Web surfing, Web-page development, and e-mail, in addition to multimedia tools. Meanwhile, schools started seriously to consider the social and moral impacts of information technology as the Internet brought about many social, cultural, legal, and moral issues.

At the start of the new millennium, the focus shifted to helping teachers and students better use computer technology. How to help students gain functional information and technology literacy, integrate technology in their learning in and out of school, and thus prepare their competency for working in an information society are hot issues both in research and practice. Various technology innovations have been introduced to schools, a variety of "good uses" of technology have been suggested to decision makers and teachers, and great efforts have been put into promoting technology use in schools. These efforts, in turn, further propel the ubiquitous diffusion of technology in schools.

Ubiquitous computing: how did it start?

An African proverb says, "Until the lions have their historians, tales of the hunt will always glorify the hunter." There are likely to be many possible stories in telling the history of one-to-one computing in education. As is frequently the case, the exact origin of an idea is often difficult to trace, and people involved in driving policy and participating in one-to-one initiatives today may draw on various historical and contemporary sources in providing a rationale for their particular initiative. One story, written by Bob Johnstone (2003a), was published in *The Age*, an Australian information technology magazine:

> The rise of one-child, one-computer learning is an extraordinary story, one in which every Australian can and should take pride. It begins in 1990. David Loader, the visionary, then principal of Methodist Ladies College in Melbourne, mandated parent-purchased notebooks, initially for years 5 and 7, and subsequently for the rest of the school. This initiative was soon replicated by forward-thinking schools such as Melbourne's Trinity Grammar and Brisbane's John Paul College. In 1995, Ken Rowe, the then principal of Frankston High School, in south-eastern Victoria, showed it was possible for schools in the public system to run successful notebook programs too. That year, Australian activities came to the attention of Microsoft executives visiting from the U.S. They were impressed by what 11-year-olds were doing with Office software in the classroom. The corporation flew 10 U.S. educators to Australia for a look-see. The Americans were stunned. "If we don't do this," one astonished principal concluded, "our kids'll be making tea for the Australians." To help spread the good word, Microsoft invited five Australian educators to Seattle. There, they were the only speakers at a conference attended by representatives from 250 U.S. schools. Thus began what became known as the Anytime Anywhere Learning initiative. By 2000, at least a thousand U.S. schools were implementing notebook programmes in emulation of the Australian model. That year, 28 schools in Britain set up a similar programme. But the most significant development was the decision by the governor of the U.S. state of Maine, Angus King, to equip every year 7 and 8 student in his state with a notebook computer.

There are doubtless other versions of this story of the origin of one-to-one computing to be told; nevertheless, this version provides a number of important insights into understanding the appeal of one-to-one computing, recognizing the power of social networks in the diffusion of innovation, and framing technological innovation in education within the wider context of economic development and globalization. First, the "one child, one computer learning" idea appears to have been initiated from within a school context, driven largely by the foresight of a principal who mandated particular policies at the school level. Second, this story, as told by Johnstone, is also about the spread or diffusion of an idea from Australia to the United States. The diffusion of the idea clearly had corporate backing and appears to have contributed to helping start the anywhere, anytime learning (AAL) initiative within Microsoft. Third, the U.S. educators' concern that their students "would end up making tea for the Australians" crystallizes a number of important ideas in terms of the perceived importance of technology in giving a nation's students a competitive edge in an increasingly competitive global economy and flattening world (Friedman, 2005).

Conclusion: ubiquitous computing in schools cannot be resisted

Ubiquitous computing in schools is undoubtedly an expensive undertaking, with complicated and unforeseen impact on teaching and learning. Constrained by budget cuts and other challenges such as meeting state standards, school leaders may wonder if they can simply resist ubiquitous computing. The answer to this question, however, probably is "no." Here, we are not advocating ubiquitous computing for all schools, but simply reporting our conclusion after considering the historical pattern of technology evolution, the current trend in ICT development, and the history of computers and education. Technology has been with human beings since the dawn of civilization, and schools have been trying to take advantage of contemporary technology since the early days of school education. By nature, the current development of ubiquitous computing in schools is not something new, but a part of this continuous endeavor.

In this chapter, we have situated the current drive for ubiquitous computing within a multidimensional historical context. Within this context, we have reviewed the evolutionary process of technology from innovations to appliances, factors that influence the adoption of new technology, the trends of the ICT development, and the impact of these trends on technology integration in schools. A short-term view of technological innovation tends to leave us caught up and mesmerized by the latest technological innovation. A historical perspective helps us to understand that computer technology has experienced an ownership pattern change from group ownership to individual ownership, an increase in the variety of size and model, changes in its functions, the distribution of expertise, and the expansion of social capital. All these developments

have propelled information technology to become increasingly ubiquitous, penetrating into our daily lives, which has significant implications
for technology integration in schools. This historical review provides
valuable perspective on the way in which technologies are implicated
in wider social and cultural changes over time.

Attentive to the multidimensional nature of new one-to-one technologies, we turn in chapter 2 to the various arguments that have been
put forward in favor of one-to-one computing.

Arguments for one-to-one computing 2

The most profound technologies are those that disappear. They weave themselves into the fabric of everyday life until they are indistinguishable from it. Consider writing, perhaps the first information technology: The ability to capture a symbolic representation of spoken language for long-term storage freed information from the limits of individual memory. Today this technology is ubiquitous in industrialized countries. Not only do books, magazines and newspapers convey written information, but so do street signs, billboards, shop signs and even graffiti. Candy wrappers are covered in writing. The constant background presence of these products of "literacy technology" does not require active attention, but the information to be conveyed is ready for use at a glance. It is difficult to imagine modern life otherwise.

Mark Weiser, *Scientific American*, 1991, p. 94

The case for one-to-one computing: merits of arguments and evidence

Underpinning the widespread one-to-one computing phenomenon are various arguments for the benefits it does or may bring. These arguments, from different perspectives and by different stakeholders, have their own strengths and limitations. This chapter introduces six arguments for one-to-one computing and examines their plausibility (see Table 2.1). Subsequently, as the combined impact of all these arguments is leading to calls for greater investment in one-to-one computing around the globe, we discuss its costs at different levels, including

Table 2.1   The arguments: moving toward one-to-one computing

Argument	Key questions	Relevant evidence
The fear argument: do not get left behind	Will schools, students and countries without one-to-one access get left behind socially, economically, and academically?	Links between digital literacy and other competences such as reading, mathematical, and scientific literacies in terms of economic development
The hope argument: better tools, better schools, better kids	Will one-to-one computing ensure better schools and better learners and deliver better tools?	The impact of computing on social and academic learning: blogging, social networking, and achievement
The simple access argument: learning on the bus	Will mere access to one-to-one computing result in digitally literate learners?	Levels of access and ways of conceptualizing access
The sophisticated argument: digitized field trips for twenty-first century learning	Will one-to-one access create new and more powerful modes of learning?	The impact of computer-supported learning environments and simulation software. The emergence and impact of "new literacies"
The equity argument: play fair	Will one-to-one computing lead to greater equity and social inclusion?	Initiatives to promote equity in access to computing. The learning and sociocultural dynamics of equity in schools
The aesthetic and durability arguments: They are so beautiful and rugged	What roles do aesthetics and durability play in efforts to advance one-to-one computing?	The increasing attention to aesthetic and durability aspects of computer design

the costs of bringing the computer to the children, expenses for making it usable, and other opportunity costs.

As we noted in chapter 1, from a historical perspective, there has been, to some extent, a quite logical progression in the advancement of information and communication technology (ICT) policies in education, with three waves of national e-learning policies evident by 2005 in developed countries (Conway, Zhao, & Lei, 2005; Zhao, Lei, & Conway, 2006). During the first wave in the early to mid-1990s, many countries focused primarily on getting hardware and software into schools as well as upgrading teachers' skills to use the new ICTs. A second wave, in the late 1990s, focused on integrating ICTs into teachers' everyday teaching. A third wave, attentive to the ways in which subject cultures and school organization often inhibited the integration of ICTs in everyday teaching, has begun to pay more attention to how school cultures, with their history of deeply embedded practices involving older technologies (e.g., pencils, slide projectors, overhead projectors, television/video) and subject matter, can more effectively integrate digital tools into the daily fabric of teaching and learning and use digital tools to change existing school practices.

Contemporary arguments for one-to-one computing occur in a situation where it is already a reality for a minority of students in some sections of education in most, if not all, developed countries. However, this partial, relatively small-scale move toward one-to-one computer provision in schools is taking place in societies where access to and use of computers for most students is still rather limited, and where there are significant differences among schools and countries in students' access to Internet computers. For example, a European Union (EU) report on Internet computers per 100 students in K–12 education across the EU identified significant disparities between countries. For example, according to 2004 data, Denmark, the Netherlands, and Norway had 26.3, 20.0, and 22.7 per 100 students, respectively, whereas Germany, Greece, and Ireland had 7.7, 5.9, and 8.7, respectively. Surprisingly, the equivalent U.S. figure, based on 2003 data from the U.S. Department of Education, was 4.4 (U.S. Dept. of Education, 2005), whereas the EU (25 countries) average was 9.9 Internet computers per 100 students (European Commission, 2006).

The arguments for one-to-one computing reflect a variety of motives and echo the widely used arguments presented in educational discourse on the imperative of enhancing ICT use in education: not getting left behind; creating better schools through better tools; learning anywhere and anytime; digitizing field trips for twenty-first century learning; creating more equitable educational opportunities for all students; and keeping up with other schools, districts, regions, and countries. The widespread one-to-one computing phenomenon moves the debate from merely getting computers into schools and classrooms to giving each child his or her own digital tools.

Of central concern to us in this chapter is the manner in which advocates of investment in computers in education—initially in terms of computer access for children at the group level, but now in terms of one-to-one access—have appropriated society's core educational goals. That is, it is argued that investment in and integration of learning technologies in classrooms, and especially advancement of one-to-one computing, will support the twin goals of social and economic development. Each of the arguments we outline in the rest of this chapter addresses one or both of these important outcomes of schooling.

The fear argument: getting left behind

Ireland lags significantly behind its European partners in the integration of information and communication technologies (ICTs) into first- and second-level education. The need to integrate technology into teaching and learning right across the curriculum is a major national challenge, which must be met in the interests of Ireland's future economic well being.

Department of Education, Ireland, Schools IT 2000, 1997

The fear argument is relatively straightforward: Fail to plan for computers in schools, plan to fail economically. Since the 1990s, when the emphasis was on securing access to computers for groups of children, and now in the current push toward one-to-one computing, the argument focuses on warning what would happen to countries and schools that did not focus investment on new information and communication technologies. No words are minced: Schools, nations and students will be left behind socially and economically. Those that do not invest sufficiently in high-quality computing will be left outside the door of the knowledge society. Typically, a failure to invest in computing in education is linked with projected demise in science, technology, engineering, and mathematics (STEM). That is, unless nations, schools, and students adopt a laser-like focus on computing to ensure access to the highest quality computing that is possible—that is, one-to-one access—economic decline is inevitable.

Thus, the fear argument warns that, unless schools of the future make computing central to students' educational experiences, they will not deliver the kinds of academic and social skills essential to prepare students for a twenty-first century knowledge-driven society. Societies that fail to deliver tech-savvy school graduates will ensure their nation's demise and marginalization from the "network society" (Castells, 1997). That is, the computer-supported "time–space compression" that characterizes today's globalizing networked world, in which social and economic transactions are increasingly undertaken in cyberspace, ensures that nations and schools that fail to invest in educational technologies will bring about their own declines.

For example, whether it is arguing for the development of e-learning in general or investing in one-to-one computing, the preceding quotation from Ireland's 1997 landmark ICT policy document Schools IT 2000 typifies the fear of getting left behind other countries that frequently underpins, in very significant ways, the drive to invest in ICTs. Furthermore, Schools IT 2000 also cited research from International Data Corporation (IDC) that ranked Ireland 23rd "in terms of state of preparedness for the information age" (para. 2.1). Not lagging behind by investing in ICTs is a powerful argument, at the local school level or at national and international levels. The pressure not to get left behind or not to lag behind has been fuelled in the wider education landscape by the use of performance rankings, which provide a very powerful rhetorical tool for those with a change, investment, or reform agenda in education (Jaworski & Phillips, 1999). Thus, in the Irish case, Ireland's 23rd place on the IDC's ranking was used as the initial justification for investment in ICTs and e-learning.

This example points to a wider and immensely influential trend in education across many countries; that is, over the last 20 years, ratings and rankings have been widely used to assess public service performance. School and university rankings, local authority or district league tables, microlocal crime data, and health trust ratings are now part of our lives. These rankings feed into policy debates at national and international levels and are often used as levers to direct attention at perceived inadequacies of countries' performance. In the area of ICTs and education, such rankings and comparative indicators are now part of the fabric of policy discourse and a low rank on some education-related ICT international comparative measure gives powerful ammunition to those with an axe to grind on a particular e-learning issue.

For example, in January 2006, eSchool News Online published an article titled "Major Study to Probe Ubiquitous Computing: Findings Likely to Impact Future Educational Technology Decisions."[1] Notwithstanding the article's conflation of second- and third-paradigm computing, the rationale for the 2006 (see www.ads2006.org) nationwide study in the United States of one-to-one computing typifies the fear of not wanting to be left behind in the international computing derby. Although no evidence is cited, the article claims that one of the consulting firms undertaking the nationwide study commented: "Hayes said she hopes the findings will be used to influence future policy decisions and initiate actions to help bridge the widening achievement gap

U.S. students now face when compared with their counterparts in other industrialized nations."

The article goes on to describe the background to the survey, including the publication of a joint preparatory report, "America's Digital Schools 2006: A Five-Year Forecast," published by the two consulting firms involved in the nationwide survey. According to *eSchool News Online,* findings from that report indicate that 87% of districts surveyed in the United States in 2005 said no one-to-one program existed in their district. Thus, even though 33 states were involved in one-to-one laptop programs in 2005 (see chapter 1), 87% of districts in 2005 reported that they did not have a one-to-one program.

These two sets of figures present different story lines. The widespread uptake of the one-to-one concept in 33 states speaks of its widespread appeal; the more cautious uptake represented by only 13% of districts nationally reporting involvement in one-to-one programs speaks of the embryonic status of one-to-one computing at a system-wide level to date. However, the initial use of the fear argument—that the United States was being left behind by other countries in the one-to-one computing derby–provided an initial discursive space for other arguments to be addressed. We now turn to some of these other arguments.

A variation of the fear argument, especially pertinent at the local level for schools, is the business argument, or "the Joneses already have it." In the opening quotation in the preface to this book, Papert's parable identifies the success of the "pencil," the new ICT, prompting a question as to the potential of ICTs in school. The next step in this scenario is that, when one school gets the latest digital tool, all other schools look over their shoulder and ask, "Why don't we invest in these tools?"—not only for the sake of learning in school, but also for the symbolic status appeal of having society's latest digital tools "in our school building" (Cuban, 2001). The equation's economic growth logic is seductive: Increased human capital (in which ICT literacy is critical) leads to enhanced economic prosperity. The fear of not getting left behind internationally at an economic and educational level is an argument we have already noted.

"The Joneses already have it" is a more local version of this argument, and it highlights the way in which individual schools are increasingly under pressure, in an era of educational marketization, to prove that they are up to date, up to speed, and on track for the twenty-first century. As Cuban (2001) argues, few school principals, superintendents, or boards would argue against the adoption of the latest wave of digital technology. Increasingly, as the appeal of one-to-one computing grows, so too will pressure on schools and school systems to join the trend—due also to decreasing costs, the accumulation of evidence that laptop programs appear to "make a difference," skillful marketing by IT companies, and the increasingly tech-savvy students already familiar with the latest multimedia technologies on their 3-G cellular and mobile phones and other digital devices to which they have access at home.

To what extent does this fear argument make sense? Will nations, schools, and school graduates be left behind unless one-to-one computing is central to education? At one level, the fear argument seems naïve since it could be argued that other educational outcomes, such as numeracy and literacy, are even more essential to nations, schools, and school graduates thriving economically and socially. Despite the alleged importance of computing, Organization for Economic Cooperation and Development (OECD) governments identified four essential literacies for twenty-first century living: reading literacy, mathematical literacy, scientific literacy, and problem-solving literacy (OECD, 2003). This is not to argue that these governments have overlooked computing in education or think it is not important; nevertheless, ministries of education, trade, enterprise, and finance are keenly attuned to the balance or blend of competences needed in "high-skill" societies. They are acutely aware that, for example, citizens without reading literacy will be even more disadvantaged than those that do not have IT skills. What is true for reading, mathematical, and scientific literacies could also be said for one-to-one digital literacy: It is important, maybe even essential, but by itself insufficient.

The hope argument: better tools, better schools, better kids

A corollary of the fear argument is the hope argument: ICTs, and especially one-to-one computing, will bring highly sought-after outcomes to education, such as higher standards, better ways of thinking, more thoughtful learners, and more powerful learning environments in schools. That is, successful adoption of one-to-one computing will catapult the school and the nation toward economic and educational success, surpassing its neighbors in a flattening world (Friedman, 2005). Essentially, the hope argument is about learning. It proposes that students will learn better in schools through enhanced access to computers in one-to-one technology-based learning environments. The hope argument is especially appealing today when educational reform has increasingly come to mean better learning (Hubbard, Mehan, & Stein, 2006). If educational reform is about learning, and computers, it is argued, are the best way to enhance learning, then one-to-one computing is the great new hope.

The arguments advanced for one-to-one computing based on the statewide laptops initiative in Maine are a good example of the hope argument. Headlines and report titles for the Maine laptops initiative (and others) on Apple's 1 to 1 Learning Web site point to some of the hopes (http://www.apple.com/education/k12/onetoone/research.html):

- Finding Proof of Learning in a One-to-One Computing Classroom (Maine)
- Laptops Unleashed: A High School Experience (San Francisco)
- Maine Evaluation Report Validates Improvement in Student Learning (Maine, Center for Educational Policy, Applied Research and Evaluation, University of Southern Maine, 2004)
- Canada's 1 to 1 iBook Program Raises Student Scores (Canada, Horizon Research and Evaluation, Inc., 2003)
- iBooks in Maine High School Provide Positive Results (Maine, Senator George J. Mitchell Scholarship Research Institute, 2006)
- Proven Academic Success (The Laptop Program Research, Rockman et al., 1997, 1998, 2000)
- Laptop Learning: A Comparison of Teaching and Learning in Upper Elementary Classrooms Equipped With Shared Carts of Laptops and Permanent 1:1 Laptops (Russell, Bebell, & Higgins, 2004)

All of these studies purport to provide evidence attesting to the value of laptops as a way of enhancing each student's learning and thereby the capacity of the school to ensure high-quality education for its community. Furthermore, the merits of one-to-one computing in terms of mobility and flexibility—the "better tool" claim—are frequently touted as evidence that these qualities enable students to interact with computers in a more personal, engaged, and enduring fashion, and so have an impact on student learning. In summary, three hope arguments are particularly influential: better tools, better schools, and better kids (in terms of achievement scores in key curricular areas such as literacy and mathematics).

How are we to evaluate the hope argument? As we have asserted here, the hope argument is about learning. To what extent do computers in general and one-to-one computing in particular enhance learning for students? Much has been written about the potential of computers to benefit student learning, but there is comparatively little research to support the claim that computers actually do enhance student learning. This is the case despite the extensive investment in ICTs and educational computing over the last decade. However, this is not to say there is no research with which we can address these questions to appraise the validity of the hope argument vis-à-vis one-to-one computing.

The better schools argument focuses on the way in which laptops and other personal and mobile digital tools can enhance a school's learning environment by enabling a more interactive, constructivist-compatible approach to teaching, as well as the promotion of more inclusive classrooms and schools. Table 2.2 summarizes key goals from a number of laptop or one-to-one computer projects in terms of the measures typically of interest to policy makers and practitioners. Perceived success on these measures has led to an increased interest in the potential of one-to-one computing since the late 1990s. However, the results are mixed. Reviewing Table 2.2 and Table 2.3, we note the following three important observations about one-to-one computing initiatives. First, there has been a distinct bias toward literacy rather than mathematics, science, or other school subjects in the focus of laptop-based, one-to-one computing initiatives. This is consistent with Australian researchers Lankshear and Bigum's (1999) observation that literacy has been the predominant focus of efforts to integrate technology into the curriculum. Consequently, while there is now some evidence to support the use of one-to-one computing, there is a lack of evidence to date in, for example, mathematics in terms of the potential of one-to-one computing to impact learning processes and outcomes.

Second, the impact of laptop-based one-to-one computing is most evident in students' writing. A range of studies point to enhanced writing according to a variety of measures: standardized tests, grade point average, the amount and quality of writing (we return to the meaning and challenge of evaluating effects, impacts and outcomes in chapter 7). Third, there is some evidence to suggest that one-to-one computing helps create the conditions for more interactive teaching, with greater student choice and more opportunities for feedback by teacher to students (Rockman, 1999).

The greater opportunity for more formative feedback is an important issue, given the increasing attention to formative assessment drawing heavily on Black and Wiliam's landmark study, *Inside the Black Box* (1998). We return to the potential of ICTs, and especially one-to-one

Table 2.2 Laptop project evaluation outcomes in literacy, mathematics, motivation, and approach to teaching

Project, location, and grade level	Literacy	Mathematics	Student motivation	Teaching
Rockman Longitudinal Study, United States	Improvements in writing	NA	Students more collaborative and involved in school work	Teachers adopt more interactive teaching methods
Canada's iBook Wireless Writing Program, grades 6 and 7	Improvements in writing on standardized measures and decrease in gender gap	NA	Enhanced student motivation in terms of work habits, attitude, and organization	Teachers provided more choice for students and used more feedback
Comparison study: one-to-one and shared mobile carts, New Hampshire, grade 7 in state's poorest schools	Greater improvements in writing in one-to-one classes (i.e., more and better writing)	NA	Greater enhancement of student motivation in one-to-one classrooms	Greater integration of computing in all subjects in one-to-one classes
Two year study of 10 laptop schools in California and Maine (Warschauer, 2005/2006; 2006)	Improvement in students' writing composition. Fostered students information literacy skills, (i.e., twenty-first century skills)	NA	Greater engagement with multimedia	Greater integration of computing in all subjects in one-to-one classes
Longitudinal study of laptop immersion program, Harvest Park Middle School, CA, grades 5–8	Improvements in writing and reading on GPA and state-normed tests in both cross-sectional and longitudinal analyses	Improvements in math on GPA and state-normed tests in both cross-sectional and longitudinal analyses	NA	NA

Table 2.3   Access to devices, conduits and literacy: measures and diffusion

	Devices	Conduits	Literacy
Focus	Physical presence of devices in specified contexts such as home, school, work, region, or country	Number connected and nature of the connection (e.g., dial up vs. broadband)	Meaningful engagement in various types of literacy practices (e.g., functional, civic/political) and specific domains of competence (e.g., knowledge, skills, and dispositions of learners). Literacy may be linked to specific technologies (e.g., digital literacy)
Key measures	Density of device in given context (e.g., Eurostar Community Survey, 2004, documented PC penetration in homes at 79% [Denmark], 50% [France], 46% [Ireland], and 36% [Poland])	Percentage connected and quality of the connection	Multiple measures encompassing types of literacy (basic vs. advanced) focusing on knowledge, skills, dispositions, and levels of competence
Speed of diffusion	Very fast/fast	Slow/moderate	Slow

computing initiatives, to create more opportunities for feedback in a later chapter.

The simple access argument: learning on the bus

> We are going to demonstrate the power of one-to-one computer access in a way that will transform education.
>
> **Former Maine Governor Angus Kinge**

The simple access argument runs as follows: Digital tools will transform education. The potential of ICTs in general and one-to-one computing in particular to transform education has been the focus of considerable debate over the last decade. Oppenheimer, for example, has decried the manner in which education's romance with new digital tools is resulting in what he views as a shallow educational experience for students. Oppenheimer tends to concentrate on the inherent weaknesses in the digital tools themselves and the manner of their use (see *The Computer Delusion*, 1997) and how, despite the highblown rhetoric, they are most likely to foster "flickering minds" (the title of his 2003 book) rather than deep and authentic learning. In his book, *Oversold and Underused: Computers in the Classroom* (2001), Cuban explains the failure of computers in schools to live up to their promise in terms of the powerful enveloping effect of schools' existing organizational cultures, the result of deeply embedded practices such as the 50-minute lesson, the division of knowledge into subjects, the persistence of the nineteenth century factory school model in the twenty-first century, and various accountability mechanisms overwhelming any e-learning innovations. The title of one of Cuban's articles succinctly summarizes his views: "Computer Meets Classroom: Classroom Wins!" (1993).

Both Cuban and Oppenheimer would agree that the digital tools movement in education promises more than it can deliver. However, they provide markedly different reasons for the failure of digital tools to live up to their promise: Oppenheimer focuses on the inherent weaknesses of the tools themselves, and Cuban on the "grammar of schooling," which means that existing cultural and organizational arrangements in schools trump any hand played by would-be e-learning innovators in school, thwarting well-meaning and well-planned e-learning initiatives (Cuban, 2006). Both Oppenheimer and Cuban would be highly skeptical of the former Maine governor's preceding claim that Maine would demonstrate how one-to-one computing will transform education.

The often simplistic arguments about how the last decade's digital technologies will transform students' learning experiences at school and at home has been a common feature of e-learning policy discourse. For example, the Texas Education Agency's 1997 state technology plan typified some of the simple rhetoric that abounds in the promotion of both e-learning and one-to-one computing. While the focus of the state technology plan at that time was making a case for investment in hardware, software and teacher upskilling to use computers in schools, the tone captures very well the belief in technology as some sort of *deus ex machina* that will redress generations-old societal inequities and rewrite the grammar of schooling rooted in nineteenth-century conceptions of the school as factory and the learner as a blank slate.

Imagine a home …

… where every parent regardless of native language or socio-economic background can communicate readily with teachers about children's progress, improve parenting skills, and get a degree or job training without leaving home or work.

Imagine a school …

… where every student regardless of zip code, economic level, age, race or ethnicity, or ability or disability can be immersed in the sights, sounds, and languages of other countries; visit museums; research knowledge webs from the holdings of dispersed libraries; and explore the inner workings of cells from inside the cell or the cold distance of outer space from inside a virtual spacesuit.

As Zhao and Conway (2001) noted in their review of 15 state technology plans in the United States:

The seductive image painted in the above excerpts [i.e., quotes from the Texas Education Agency] sets in motion a sales pitch, typical of the other state educational technology plans we reviewed. Each, in similar ways, was trying to sell technology by projecting a tempting vision. Each, in trying to sell technology, skated lightly over any need to present research about the intricacies of meeting the promised land or outcomes of such multi-million dollar investments. Each, in trying to win customers, relied more on exclamation about the benign nature of technology, singular, rather than explanation about the constraints and possibilities of various technologies, plural. Each, in trying [to] sell a politically fair plan, relied more on sloganizing about equity than elaborating on ways of redistributing resources in favor of those traditionally marginalized in past waves of technology innovations in schools. Each focused more on future possibilities than present constraints and past failures.

A second dimension of the simple argument focuses on access and runs as follows: Get these new digital tools into the hands of individual students so that they can use them whether they are at home or school. At one level, this is not problematic. Clearly access matters; it is hard to use a tool without actually having access to it. However, the focus on solving the individual access problem pushes the equally, if not more, important issue of the one-to-one program's social context into the background. Simple access arguments win out over context sensitivity. Distinguishing between what Warschauer (2003) calls access to devices, technology as conduit and digital literacy provide a more comprehensive framework for thinking about one-to-one digital tools in relation to educational goals and the social context of their use in homes and schools.

Warschauer criticizes simplistic policy emphases that only focus on access as measured by density of devices in homes or schools (x number of computers or laptops per student, per school, per home). The second part of this access to devices logic is a focus on fast diffusion. Many policy makers and technology advocates in business or academia often use and are easily won over by statistics demonstrating rapid rise in diffusion of specific devices. The technology-as-conduit argument challenges the access-to-devices logic by asking whether and how those with access to devices are connected. As Warschauer notes, "Whereas a device can be acquired through a one-time purchase, access to a conduit necessitates connection to a supply line that provides something on a regular basis" (2003, p. 20).

Noting that radio and TV are quintessential conduit services in that "they are worthless without the accompanying airwaves" (2003, p. 24), Warschauer argues that diffusion of conduits has historically been a lot slower than that for devices across many cultures, largely because "diffusion of technology is a site of struggle, with access policy reflecting broader issues of political, social and economic power" (p. 25). The various paths that electrification has taken around the world, he argues, testify to this manner in which conduit access is a deeply political matter. For example, the emphasis on electricity as a social service rather than a private commodity in Europe, as compared to the United States led to somewhat different combinations of market forces and government actions to ensure universal access to electricity and the various technologies that could be built upon an electrical infrastructure. However, Warschauer argues that neither the access to devices nor conduits categories fully capture the essence of "meaningful access to information and communication technologies."

What is most important about ICT is not so much the availability of the computing device or the Internet line, but rather people's ability to make use of that device and line to engage in "meaningful social practices" (Warschauer, 2003, p. 20). Warschauer's sophisticated argument points to the difficulties of reaching the high watermark stipulated in e-learning policy aspirations.

The sophisticated argument: digitizing field trips for twenty-first century learning

Perhaps the strongest educational argument advanced in favor of one-to-one computing is that it will prepare students with the skills they need for twenty-first century life as consumers, workers, and citizens. The potential mobile and personal affordability of one-to-one computing are seen as the basis to providing long-term, high-quality meaningful engagement within powerful learning environments. For example, Zurita and Nussbaum (2004), in their study of handheld computers in Chile, noted how computers connected to a network facilitate collaborative learning and types of learning that could not occur otherwise. Others argue that using networked computers can also engage learners with sophisticated content that is otherwise difficult to teach through simulations of traffic patterns, population dynamics, or other ideas difficult to explain in the absence of dynamic visual forms of representation (Colella, 2000; Klopfer & Yoon, 2005; Klopfer, Yoon, & Rivas, 2004; Roschelle, Pea, Hoadley, Gordin, & Means, 2000). Continuous access to wireless computers also provides students, it is argued, opportunities to search for, evaluate, synthesize, and communicate knowledge to others and as such constitutes a powerful reason why one-to-one computing is essential in that it prepares students to engage with information and ideas in ways that are and will continue to be a part of the twenty-first century (Florida, 2000; Rockman, 2003; Warschauer, 2006).

A key feature of Papert's story that we quoted in the preface to this book is the way in which the pencils, as the new ICT in his parable, were attractive to leaders in business and science and the integration of the new technology was eventually identified as appealing in preparing students for life. In a similar fashion, the use of desktops, laptops, and other personal digital tools is increasingly seen as essential to equip students with the skills they will need for the lives they will lead. After over a decade investigating the use of laptops in schools, Rockman concluded that one of the most significant benefits of laptop initiatives is their capacity to increase so-called twenty-first century skills:

> Developing the ability to learn independently, collaborate with peers to accomplish work, and communicate the conclusions of your work are the core of 21st century skills, and a highly valued set of competencies in the world outside of school. These accomplishments are seen in many laptop programs, especially those

that permit students to take their computer home in the evening. (Rockman, 2003)

According to this argument, the sustained, well-integrated use of laptops promises to create contexts within which schools can educate students who are skilled consumers and creators of information, have good communication skills, can engage in thinking and problem-solving, and have the capacity for good interpersonal engagement.

The one-to-one computing phenomenon is part of a wider change, especially in developed countries, where the discourse about developing the knowledge economy has come to dominate and shape debate across a range of public policy areas, and especially in education (Hargreaves, 2003). Discussions about the knowledge economy often identify e-learning as both a critical tool and a goal: a tool in preparing students to work in a digital-rich world and a goal in that failure to capitalize on cutting-edge digital tools in schools, in the workplace, and in wider society will ensure that neither a society nor its citizens reach their potential or are competitive in a flattening world (Friedman, 2005).

To what extent do the claims of the digitizing field trip make sense? The most convincing arguments supporting this position focus on: (1) opportunities to engage in powerful learning environments; (2) the development of twenty-first century skills through engaging in such learning environments over time; and (3) the promise of ever more powerful and mobile connectivity making such learning environments increasingly viable. On the other hand, the very high cost of developing software and learning environments to support extensive curricular content across many different school subjects is an enormous challenge, despite compelling "boutique examples" of digitized field trips.

Let us review the evidence that supports the digitized field trip argument. First, one-to-one computing has the potential to connect students to powerful learning environments. Such environments are epitomized by "digitized field trips" where students can actively engage in simulations, connect with scientists and other experts, and learn key ideas in the virtual world created CSILE (Computer-Supported Intentional Learning Environments), Scardamelia & Bereiter, 2000). Typically, a model of what constitutes a powerful learning environment is central to this sophisticated understanding of the potential of one-to-one computing.

For example, Roschelle et al. (2000) identify several examples of computer-based applications to illustrate ways technology can enhance how children learn by supporting four fundamental characteristics of learning: (1) active engagement, (2) participation in groups, (3) frequent interaction and feedback, and (4) connections to real-world contexts. Thus, the argument advanced in this instance takes account of the conditions under which optimal learning with digital technologies is likely to occur. To the extent that students can gain regular and reliable access to powerful learning environments, one-to-one computing provides an important new context beyond conventional face-to-face learning opportunities. Increasingly, as computers become more reliable and more powerful, learners may have to work within such environments over a long period. This opportunity for extended engagement in such environments is a compelling argument in favor of one-to-one computing because up to now many of the most powerful applications of technology-enhanced learning (e.g., simulations; see Roschelle et al., 2000) have been available only to a small number of learners.

Second, evaluations of some one-to-one laptop initiatives have argued that they appear to support the development of information literacy skills such as analyzing, filtering, and critiquing information (Rockman, 2003; Warschauer, 2006). Warschauer (2005/2006) viewed this outcome as one of the most beneficial features of laptop initiatives studied in Maine and California. The evaluation of Microsoft's Anywhere, Anytime laptop program came to a similar conclusion, asserting that, through continuous one-to-one wireless laptops, students can learn to search for, access, filter, and organize information into useful knowledge that can be communicated to others (Rockman) and that developing these capabilities is an essential aspect in acquiring "twenty-first century skills."

Third, one-to-one computing has the potential in the coming years to provide access via more powerful broadband capacity to the sort of powerful learning environments we have been discussing (3 G at present but 4-G technologies will transfer information up to 100 times faster than 3 G by 2010 (Boston, 2003). For example, an increasingly important dimension of the digitizing field trip argument is the advent of 3-G technologies that are turning cellular and mobile phones and other devices into multimedia players, making it possible to access audio and video clips. (Three-gigabyte wireless networks are capable of transferring data at speeds of up to 384 Kbp. Average speeds for 3-G networks range between 64 and 384 Kbp. This is a significant jump from common wireless data speeds in the United States that are often slower than a 14.4-Kbp modem). Students' out-of-school experiences with 3-G technologies are raising the bar for schools in terms of the types of media being deployed by teachers. Consequently, the digitized field trip is becoming or likely to become a necessity rather than an optional extra or added bonus in order to engage today's and tomorrow's tech-savvy children and adolescents.

On the other hand, as noted earlier, developing appropriate curriculum materials for each grade level in every school subject for students of different levels of achievement and cultural background presents an enormous challenge. For example, in many parts of the world, a multiplicity of languages presents a significant challenge to the development of culturally appropriate textbook and online resources for use by teachers and students.

The equity argument: play fair

Why do children in developing nations need laptops?
Laptops are both a window and a tool: a window into the world and a tool with which to think. They are a wonderful way for all children to "learn learning" through independent interaction and exploration.

Frequently Asked Questions section of One Laptop per Child (OLPC) Web site, available online at: http://laptop.media.mit.edu/

The play fair argument arises at the local level within and between schools, at regional and national levels, and at the international level. For example, at a global level, one-to-one computing has been driven primarily by the One Laptop per Child (OLPC) initiative. The OLPC movement and numerous other projects at national levels in various countries share similar goals: the provision of one-to-one access, overcoming geographical isolation, addressing inequities within countries, and targeting social division between the haves and the have-nots. One could summarize the basic logic of the equity argument as follows: Unless each student has access to high-quality one-to-one computing, the persistent computer-to-student inequities evident within and between school systems will exacerbate social exclusion.

Efforts to address educational equity and promote social inclusion have concentrated on literacy and numeracy in many countries—in part because economic analyses of the rate of return on investment provide significant evidence that good literacy and numeracy skills act as a protective factor for students and communities deemed at risk (Wolf, 2002). Within this literacy/numeracy focus, according to Lankshear and Bigum (1999), literacy and language initiatives have become a focus of efforts to integrate technology into the curriculum (see Kamil, Intrator, & Kim, 2000; Lankshear & Bigum). Given this international trend toward planning, implementing, examining, and understanding the role of digital tools in promoting students' literacy learning, including those students with literacy difficulties, laptop initiatives have become a test-bed for the potential of one-to-one computing. These projects have provided some evidence—by no means unequivocal—that one-to-one computing can address wider concerns about equity.

As one-to-one computing studies and evaluations accumulate, there is considerable potential to raise questions about and influence policy on the promotion of social inclusion in the emerging knowledge society (Hargreaves, 2003). Thus, for example, for those students deemed most at risk locally, nationally, and globally because of low levels of literacy and its attendant long-term social and economic consequences, many people are arguing that the provision of one-to-one computing is imperative. Furthermore, consistent with the international trend toward creating more inclusive mainstream classroom settings for students with special needs (Evans & Lunt, 2002; UNESCO, 1994), digital tools are being promoted as key components in efforts to develop models of e-learning-based inclusion that work for schools, teachers, and students across vastly different cultural and educational contexts (UNESCO, 2003).

To what extent can technology bridge long-standing and powerful societal inequities? The appeal of technology in this context is very powerful, but it is not always supported by research evidence. For example, Warschauer (2005/2006) used the term "*Sesame Street* effect" to refer to the way in which laptops in Maine and California did not reform troubled schools and bridge achievement gaps. That is, the laptop initiative, like *Sesame Street,* tended to amplify rather than overcome achievement gaps even though, like the children's television program, it was designed as technology intervention to bridge achievement gaps and address wider societal inequities at an early stage of child development. Furthermore, as we noted earlier in relation to the simple access argument, when considering the digital, the provision of one-to-one computing hardware is meaningless unless we address the quality of connectivity and actual digital learning experiences of students.

The aesthetic and durability arguments:
they are so beautiful and rugged

As human–computer interaction (HCI) and interactive systems design have developed a sense of people living with and through technologies, our concerns have broadened from usability to include wider qualities of people's experiences with technology.

McCarthy, Wright, Wallace, and Dearden, 2005, p. 1

The aesthetic appeal and durability of recent products from companies seeking to serve one-to-one computing demands in education is an increasingly salient aspect of how one-to-one computing as a concept is being marketed. Whether it is the compact beauty of the Microsoft Tablet PC, the rugged appeal of the OLPC laptops, or the sleek and eye-catching iBook, the way in which new personal and mobile computing is inherently enchanting and not merely functional is an increasingly important aspect of human–computer interface design (McCarthy et al., 2005). The importance of durability in one-to-one computing is abundantly clear in the emphasis by the OLPC initiative on the spill-resistant keyboard and carrying handle of its laptops for students in developing countries.

The move toward more aesthetically pleasing human–computer experiences is reflected in a number of research fields, including human–computer interface (HCI) research, research on human motivation and computers (e.g., the potential of computers in flow experiences), and studies being undertaken in the rapidly growing field of personal and ubiquitous computing examining such issues as "calm computing." With the increasing appeal and growth of one-to-one computing, understanding the subjective experiences of users has become a common theme across these different areas of research. We focus on one line of research here to illustrate the aesthetic argument using the concept of "flow" in relation to one-to-one computing.

"Flow" as a concept has been developed most especially through the work of Csikszentmihalyi as a means of understanding the psychology of optimal experience—that is, moments of intense engagement that are truly enjoyable for participants, whatever the nature of the flow-inducing activity may be. As Csikszentmihalyi notes:

Watching a good play or reading a stimulating book also seems to produce the same mental state. I called it *flow*, because this was a metaphor several respondents gave for how it felt when their experience was most enjoyable—it was like being carried away by a current, everything moving smoothly without effort. Contrary to expectation, *flow* usually happens not during relaxing moments of leisure and entertainment, but rather when we are actively involved in a difficult enterprise, in a task that stretches our mental and physical abilities. Any activity can do it. Working on a challenging job, riding the crest of a tremendous wave, and teaching one's child the letters of the alphabet are the kinds of experiences that focus our whole being in a harmonious rush of energy, and lift us out of the anxieties and boredom that characterize so much of everyday life. (1991, p. 12)

Csikszentmihalyi notes some key preconditions for experiencing flow, including the observation that it is more likely in demanding leisure and work activities than in relaxing activities such as watching television. In the latter instance, Csikszentmihalyi notes that people report higher levels of stress, depression, and tension after watching television than in relation to other, more demanding leisure activities. Thus, flow experiences are more difficult to achieve in the absence of effort and active mental or physical engagement in an activity. Activities that include clear, achievable goals and immediate feedback are considerably more likely to induce flow. Of particular importance in the context of one-to-one computing and evidence that computer users can become intensely engaged in using computers, to the point of addiction, is that some people are particularly adept at setting goals that increase the likelihood of flow even in what objectively seem unlikely occasions for flow-like experiences. Finally, flow tends to produce some personal growth through the sustained engagement in the activity since the experience of flow brings individuals back for new challenges and a growing sense of personal skill levels or computer efficacy in the case of one-to-one computing (Csikszentmihalyi, 1991).

   In relation to one-to-one computing, there is a strong case to be made that the potential for many different levels of engagement with computers, while not guaranteeing flow, at least makes it more likely to provide opportunities for flow than the generally more passive and challenge-free activity of watching television. For example, Hoffman, Novak, and Duhachek (2002) demonstrated that, contrary to some previous studies, flow is more likely to occur in task-oriented online activities than during recreational activities.

   How can we understand the aesthetic and durability arguments within the context of the current debate on one-to-one computing? Take, for example, the OLPC initiative, which has made both aesthetics and durability key components of the design. Rather than afterthoughts, they are seen as part of the full one-to-one computing experience. Thus, what may seem like less important factors than others we have addressed, beauty even more than durability, are being considered as central in human–computer interaction. They are important dimensions of the one-to-one computing phenomenon because they signal the value of the quality of the student–computer experience as critical in framing learning experiences.

But can we afford it? Costs of one-to-one computing

Some critics have also complained that wealthy donors should concentrate more on less glamorous projects like stamping out malaria before trying to give every child an e-mail address. But Negroponte insists that mass connectivity and education are the solution to nearly all the world's ills, "from poverty to peace to the environment." If he's right, the $100 laptop could be the little box that saves the world.

**Michael Crowley, *The New York Times*
December 11, 2005**

*One-to-one computing as "must-have resource of
2010": costs in an accountability culture*

The 'must-have technology' of 1853 (Varian, 2006), according to the *New York Times* (February 9, 2006), was the sewing machine. Its success was largely owing to the fact that it offered purchasers a way to make money by taking up mending. Isaac Singer capitalized on this application by inventing a new way to sell products to consumers: the installment plan. Is one-to-one computing today's must-have technology? Or are the costs prohibitive? In this section, we address the question of cost from a number of perspectives: the wider context of accountability in education, affordability, and examples of some approaches to funding laptop programs. Warschauer (2005/2006), commenting on whether such programs work, noted the importance of both levering funding from different sources and ensuring a good fit between circumstances and the funding model adopted. Singer invented a funding model that worked for the sewing machine in 1853; so, too, schools and systems are and will have to continue to be imaginative in considering various funding possibilities.

In the current educational climate, effectiveness alone is not a sufficient criterion for implementation of one-to-one computing. As the criticism of MIT's OLPC initiative referred to in the preceding quotation from the OLPC Web site demonstrates, in an era of scarce resources and moderately expensive technologies, in order for money to be devoted to one-to-one computing, the relative cost effectiveness of such expenditure as compared to other possible educational interventions must be established. Furthermore, the current potential to spend significant amounts of scarce educational funding on one-to-one computing comes at a time when school finance, in many countries, is increasingly being devolved to local school principals through site-based management initiatives, rather than as an administrative matter being undertaken at distance by a school administrator at the district, state, or national level (Caldwell & Spinks, 1988; O'Donoghue & Dimmock, 1998).

It also comes at a time when, due to an unprecedented culture of accountability, there is increasing attention on value for money in education (Levin & McEwan, 2001). The focus on accountability has directed attention to resource allocation models. In the case of educational reforms, analysis suggests that allocation of school resources is complicated, according to Miles and Darling-Hammond (1997), by structural constraints in the school system. Reform initiatives may end up being viewed by some as peripheral to the core academic program, fall foul of rigid and fragmented school schedules, and bump up against inflexible teacher or administrator role definitions. As such, the cost of assessing one-to-one computing has to be understood, like any other educational reform effort, within the ecology of school change.

Costs and affordability

No matter where decisions about expenditure on computers are being made, the issue of cost arises in a number of ways. The total cost of any investment in education can be seen as encompassing three types of costs: the upfront or actual monetary cost at a given point in time, the cost of upgrading or refurbishing technology infrastructure, and the opportunity cost of not investing in a particular educational innovation such as one-to-one computing. These three types of cost lead naturally into three different, yet related "but can we afford it" questions:

- What is the actual monetary cost of one-to-one computing today?
- What is the future cost of refreshing one-to-one computing, given periodic obsolescence?
- What is the opportunity cost of not investing in one-to-one computing?

Whichever of these questions is being addressed, it is the actual upfront monetary cost of one-to-one computing that is one of the most challenging and potentially off-putting aspects of the one-to-one computing phenomenon—whether one is talking about the developed or developing world. To take a developed world case, many different sources are used to fund one-to-one initiatives in the United States, including leasing, federal grants, state funding, reallocated funds from existing capital and operating budgets, philanthropic grants, local community support, and parent purchase. Even though initial costs decrease quite rapidly, the continual upgrading, as new waves of software become available and laptops with higher specifications are required to install the new software, puts schools under continuous financial pressure to keep up with the pace of technological innovation and has a potential knock-on effect on professional development and curriculum planning costs (Culp, Honey & Mandinach, 2005). Technology also provides a special challenge in relation to refurbishment and upgrading on a scale very different from, for example, the costs of periodic obsolescence that occur with some other school resources. Consequently, an important question arises regarding funding cycles, which may need to be multiannual, and the extent to which schools adopt a strategic approach to the problem.

One-to-one computing as a hybrid expenditure

Culp, Honey, and Mandinach see technology as a source of pressure on existing budgetary patterns since it can potentially impact many other line items, such as professional development and curriculum resources. They note:

> Because effective use of technology must be supported by significant investments in hardware, software, infrastructure, professional development, and support services, over the last decade,

we as a nation have invested more than $66 billion investment in school technology. This unprecedented level of investment in educational technology has raised expectations of legislators and the public who are now looking for returns on this investment (2005, p. 2).

They view technology as a type of "hybrid expenditure" in that it is necessary, especially if linked to wider educational reform efforts, to interweave it with expenditure on professional development and other curriculum resources. Furthermore, if one takes the initial cost and upgrading costs together as a budgetary planning task, Culp, Honey, and Mandinach (2005) identify the need for multiyear budgets to take a comprehensive account of the total cost of technology in order to optimize schools' investment in the long term.

Funding models: grants, leasing, ...

How do schools address the cost of funding one-to-one computing? We start with a U.S.-based case, the story of Alaska's Denali Borough School District, and another from the Laptops Initiative for students with dyslexia and other reading and writing difficulties in Ireland. The Alaskan district has an experimental, wireless-enabled school bus and has developed a leasing program as the basis for its one-to-one laptop initiative. In collaboration between the district and Apple Financial Services (AFS) Education Finance, all the district's high school students (and their teachers) were supplied with their own PowerBook G4 computers. In Ireland, the National Centre for Technology Education (NCTE) administered the €2.7 million allocated by the minister of education to 31 participating grade 7–12 postprimary schools.

These examples illustrate a very important factor in educational technology investment over the last decade: It has been the result of efforts both by governments and strategic initiatives from the private sector. The sheer cost of one-to-one computing puts it out of reach of many families, so schools have an important role in levering appropriate funding that fits local circumstances. The decision of a number of governments that have block-purchased laptops as part of OLPC is a good example of the government-led approach to addressing cost.

What funding models might best serve schools in neighborhoods across the socioeconomic spectrum? In an article titled "Going One-to-One," Warschauer (2005/2006) advised schools to "practice creative financing" and used some examples of how schools in Fullerton School District in California implemented a pilot laptop program in 2004–2005. The annual cost per student is U.S.$468, which includes a "three-year lease-to-own contract, educational software, a warranty, insurance, and a protective sleeve for each laptop" (p. 36). Two schools, located in middle- to high-income communities, organized a lease-to-purchase scheme and all parents took this option. In a low-income neighborhood, a school organized a leasing scheme using Title 1 funds. These examples illustrate how schools may have to and do adopt funding mechanisms that will work well within the context of a community's financial resources. As Fullerton's laptops initiative expands, Warschauer notes that parents will be asked to purchase laptops for their children and the district plans to develop financial support packages for low-income parents using state funds, federal funds, and/or financial donations from local business and industry or parent groups.

Finally, we come to the third question: What is the opportunity cost of not investing in one-to-one computing? Educators' mission is preparing students for a world we do not yet know but can only make well-informed guesses about, always operating with less than full information, so investment is always a calculated bet. If, for example, we agree with Florida's (2002) analysis that creative societies are characterized by an emphasis on nurturing talent, technology, and tolerance, then investing in one-to-one computing might seem a very smart move if that investment can lead to enhanced opportunities for student learning in the short and medium term, and in the longer term enhance a society's capacity to play, work, and innovate with technology.

If schools, educators, and the wider society view one-to-one computing as a worthwhile investment, for these or other reasons, market-driven business models, such as those adopted to create mobile phone saturation in many countries, could be used to encourage government- and development agency-led initiatives such as OLPC to ensure the provision of affordable laptops. Of course, investing in one-to-one computing involves opportunity costs. For example, some critics have argued that the OLPC initiative is a waste of scarce funds, and the money spent on the $100-priced laptops would be better spent on addressing the HIV/AIDS epidemic and malaria rather than on what some in developed countries perceive as the immense educational potential of e-learning (see *New York Times,* December 11, 2005).

Conclusion: arguing the arguments, sensible arguments

This chapter has presented and examined some arguments underpinning the one-to-one computing phenomenon. We have noted how these arguments, from different perspectives and by different stakeholders, have their own strengths and limitations. In conclusion, we might ask ourselves why one-to-one computing is becoming more appealing as a policy option. Whether it is a proposal to opt for a laptop initiative or a plan to introduce notebook tablets or PDAs, the initial appeal in all these initiatives internationally seems to be the potential of these digital tools to meet the ambitious goal of providing *personalized anytime, anywhere access to ICTs* for regular or special education students. One factor that has fuelled the appeal of such digital tools is the observation that, despite the fact that the computer-to-student ratio may be low in many schools, computers are often located in relatively inaccessible labs that have inhibited and continue to inhibit students' individual daily access and use. Consequently, laptops and other portable digital tools have been heralded as one solution to this problem of access. There are a number of key points with which we conclude this chapter.

First, the various arguments we have presented tend to coexist without one particular argument clearly and unequivocally winning the day. Thus, as enthusiasm for new waves of technology occurs within design and development circles, there is in education circles a simultaneous mix of skepticism, advocacy, and excitement at the potential of the new technology to transform schooling.

Second, as evidence accumulates about the complex ways in which digital tools act within schools as ecological systems, it challenges simple arguments, supports some of the sophisticated arguments, and moderates some of the hopeful arguments made for the potential of one-to-one computing to change classroom and school practices.

Third, there are significant research gaps in what we know about the use of one-to-one digital tools in education. There is a need for carefully designed ecological studies, for while much is said about how digital tools, in general, or one-to-one computing, in particular, enhances student motivation, we actually know very little about the nature of the motivation that e-learning inspires. For example, motivation theorists often distinguish between intrinsic and extrinsic motivation, and have developed very useful models of students' goal orientations as a way to understand learning in school settings (Pintrich & Schunk, 2001).

As yet, the supposed motivational benefits to learners of one-to-one computer use in education are largely uncharted territory.

Finally, the use of empirical studies to inform the various arguments we have noted in this chapter is sporadic. More often than not, claims are made based partly on previous research on technology use in education and partly on a mixture of speculation and aspiration about how one-to-one computing will evolve in education.

Having focused on various arguments in this chapter, we want to conclude by acknowledging how the hope argument, in particular, is an ever shifting horizon when it comes to one-to-one computing. For example, recent claims about how fuel-cell-powered batteries will prolong laptop use for a full day or more promise even greater levels of portability and flexibility in ICT use (Ward, 2006). Perhaps even more influential in the long term is the impact of children's and adolescents' experiences with wireless 3-G technology-driven cell and mobile phones, as these are setting new aesthetic, portability, durability, educational, and entertainment thresholds that schools, even if they do not seek to emulate them, will have to address in planning how to make use of technological innovation. As Prensky (2005/2006) and others have argued, many students "power down" rather than "power up" when they enter the slower, less stimulating experience of using computers in school, compared to their flow-like out-of-school experiences of one-to-one computing.

Conditions for laptop use in schools 3

[I want] a specialized computer that does all your homework so you can relax and watch TV all day!

An 11-year-old student

Introduction

In some significant way, we all share the wish expressed by an 11-year-old in response to the U.K.-based BBC "Tell us your technology idea for the future" campaign. We have long wished for a computer or any machine that would do the education job and cure all education ills. We have bought many machines and put them into classrooms and waited for miracles to happen. To our disappointment, miracles have not come as expected. We can, of course, blame technology for not being "helpful at all," as did a parent and board member at T. C. Williams High School, Virginia, which started a laptop project in 2004. But what could have happened is that the technology was never used; a student from the same school reported that his laptop "was pretty much under my bed all last year, except for one time a quarter, when it was mandatory" (Bahrampour, 2006). If it is not being used, how can it be helpful?

As noted in the previous chapter, there are many arguments for ubiquitous computing and various goals for implementing one-to-one laptop projects in schools, but there is only one way to make this ubiquitous computing idea work, which is for teachers and students to use laptops. How can laptops be used enough to have an impact and used in appropriate ways to make sure that the impact is positive?

If we observe an ecosystem in the natural world, for a species (e.g., a plant) to survive and thrive, many conditions must be present: sufficient resources—water, sunshine, and soil with the right minerals; a friendly and supportive environment, with enough space and the appropriate niche to grow; no severe competition from other species for the resources; no serious predators; sufficient interactions with other species, such as animals or birds, so that the seeds can be spread and this species can proliferate; and, more importantly, the species and the whole ecosystem co-adapting to each other and co-evolving in virtue cycles as the plant grows and thrives. Dramatic change in the environment can kill a plant overnight, and the rampant spread of a specific plant may deplete the resource and destroy the ecosystem (Lewontin, 2000).

Laptops in schools are similar to a species in a natural ecosystem. Laptops need resources—time, money, technical support, and peripheral technologies; they need to find the right niche to fit in the school system; they may compete with other species such as textbooks, computer labs, and the library; the environment needs to provide a supportive policy and culture; there should be enough interactions with people—teachers and students—so that laptop use can take root and spread; and in ways similar to the natural co-evolution process, laptops and the whole system co-adapt to each other. At different developmental stages, the laptops may need different resources and support, and the interactions with teachers, students, and the system also vary. Whether laptops can survive and thrive depends on this dynamic process of co-evolution and co-adaptation.

This makes the use of laptops, like that of many other technologies that have been put into schools, a very complex process influenced and constrained by many conditions. These conditions can be factors related to the school environment, characteristics of the school culture, the readiness and experiences of teachers and students for using technology, and the dynamics of social interactions in the school system.

All these conditions affect two major measures essential to the survival and prosperity of technology use in schools: One is the quantity of technology use (in other words, how much technology is used) and the other is the quality of technology use, which means how technology is used (Lei & Zhao, in press). For laptop use in a ubiquitous computing environment, both the quantity and quality of use are critical because, first, the laptops have to be used to have any effect and, second, to have positive effect on teaching and learning, they have to be used in meaningful ways. Without the necessary facilitating conditions, laptops can hardly function at either the quantity level or the quality level.

In this chapter we discuss the various conditions that influence the deployment of one-to-one computing projects, the use of laptops in schools, and how the roles of different conditions change as the ubiquitous computing environment evolves.

Conditions for effective laptop use

Conditions for effective laptop use can be discussed on two levels. One is the functional level—how to make laptops work, and the other is the meaningful level—how to integrate laptop use in teaching and learning in meaningful ways. For laptops to be functional, various issues need to be taken care of, such as basic technology facilities, safety and security issues, and technical support. For laptops to function at a meaningful level, merely putting functional technology into classrooms does not necessarily help integrate technology into teaching and learning. Holistic support from all aspects of the school system is critical in facilitating a virtuous cycle of changes and helping to grow a healthy ecosystem for meaningful laptop use.

Infrastructure

Infrastructural conditions include technology hardware, software, and supporting facilities that are available to teachers and students, such as computers and Internet access; availability of software; and the ease of accessibility to hardware, software, and peripheral technologies (Zhao, Pugh, Sheldon, & Byers, 2002). Infrastructure to technology use is similar to water to plants, providing critical resources for its development.

A basic condition for student technology use is access. When computers were viewed as expensive machines for experts, there could not be much student use, which was the case in the first few decades after computers were invented. For example, in the International Association for the Evaluation of Educational Achievement Computers in Education project (IEA CompEd), Pelgrum and Plomp (1996) found that, prior to 1992, there was little integration of computer use in the curriculum of schools, mainly due to the scarcity of computers available in school at that time.

One may assume that in a one-to-one computing environment, since everybody has a laptop, there should not be access problems. However, this is unfortunately not the case because, today, a computer has to rely on many other resources to function. In the school technology ecosystem, a computer is just one of the many components needed to enable its use (Lei, 2005; Zhao & Frank, 2003). Full access to a computer implies access to its full range of functions, supporting facilities, and peripheral technologies such as Internet connections, printers, projectors, and software programs. If these resources cannot be sufficiently provided, access is still an issue that constrains the use of laptops in schools.

To realize the goal of anytime, anywhere learning, reliable wireless Internet access is a key infrastructure. In Australia, the Department of Education and Training (DE&T) gave each of the 42,000 teachers in Victoria a laptop computer, but soon realized that the lack of connectivity was a significant impediment to the use of laptops in the classroom. Thus, approximately U.S.$4.5 million in state funding was earmarked to build the Wireless Networks for Schools (WiNS) program to equip all public schools with secure, high-speed, wireless access (Cisco, 2006) Conversely, an unreliable and slow wireless network was often reported as a problem. Many schools in Maine experienced trouble staying connected to the network in the early stage of one-to-one computing (Silvernail & Harris, 2003). Some teachers complained that when students visited a Web site to take a quiz or work on an exercise in class, the connection could become very slow. Schools surveyed in the U.K. Tablet PCs evaluation study conducted by BECTA (Sheehy et al., 2006) reported that unreliable and slow wireless networks limited the range of activities and undermined teachers' confidence in using these laptops in class; in schools without wireless networks, they commented on the need of wireless networks to better support classroom activities.

Another Internet-related obstacle is filters. Students often reported that they were not able to download files or visit Websites, which limited the information they could access and activities they could engage in. Some teachers suggested that if students were being given laptops, they should be given full privileges to use the computers (Rockman, 2003).

When every student has a laptop, the demand for peripheral technologies such as data projectors, CDs, memory sticks, digital cameras, and printers is increased. For students to work on multimedia projects such as making movies, digital and video cameras are needed. For teachers to use laptops to present their lectures, there must be enough computer projectors available. The ideal situation would be to equip each classroom with a projector. Some schools purchased large-screen television sets for classrooms as an alternative to more expensive data projectors (e.g., Zucker & McGhee, 2005).

It is often said that ubiquitous computing helps make a "paperless" environment. This might be true at a later stage of a laptop project. However, at the earlier stage, the consumption of paper usually goes up because people are used to reading on paper instead of on the screen and they want to print out their work. Hence, the need for printers may also increase as more people have laptops (Lei & Zhao, 2006).

In a large study that evaluates the Michigan Freedom to Learn (FTL) ubiquitous computing project, teachers also expressed the need for supporting technology equipment, particularly data acquisition tools that would allow students to conduct more inquiry-based science projects, such as "graphing calculators and probes, LEGO robotics, global positioning systems, cameras, projectors, CD burners, microscopes, data loggers, smart boards, scanners, and even computer-microscope links"(Urban-Lurain & Zhao, 2004). The lack of these facilitating technologies seriously limited the variety of tasks teachers and students could perform on their laptops.

Sometimes a little technological glitch can turn out to be a big problem, such as the outlet problem in Alpha Middle School. Located in an affluent suburban community, the Alpha Middle School was equipped with cutting-edge technology facilities that were not common in most schools. However, there was a deficiency in the classrooms: One classroom had only three outlets—one on the front wall of the class that was generally used by the teacher and two on the back wall for student use, this apparently was not sufficient for a class of 25 students, each with a laptop computer.

The battery life of laptops was about 3–5 hours at one full charge. The battery may or may not be able to last for a full school day, depending on the amount and type of usage. If a student fails to charge his or her laptop, it might be difficult to participate in some activities. The school had foreseen this problem and made certain regulations on this issue. For example, the "Laptop Project Policies and Procedures" brochure distributed to all students and their parents clearly stated: "Batteries must be fully charged at the beginning of each school day. Charge your computer every night." This brochure also provided some tips for students to maximize battery life, and students were advised to charge batteries during lunch periods.

Not surprisingly, every day there were students who forgot to charge their laptops and students who forgot to bring their chargers. Then, in class when the teacher asked students to work on their laptops, students with dead batteries would need to go to the back of the classroom to charge their batteries, and students who did not have their chargers with them would have to run to the technical support department to borrow one—there were not always enough extra chargers to circulate. Since there were only two outlets, students had to take turns to charge their laptops. It was very inconvenient for students and sometimes disruptive to the class. A teacher complained during the interview: "Why are there only three outlets? In a school like this, the outlets should have been installed on every student's desk." The problem was finally lessened by purchasing more chargers and power strips.

Similar problems were reported in other large laptop initiatives. For example, the Tablet PCs used in many schools in the United Kingdom had a battery life of about 3 hours, which was too short to last the whole school day. Therefore, it was considered vital to provide charging capacity in classrooms such as charging trolleys (Sheehy et al., 2006). The Virginia Henrico County Public Schools (HCPS) also reported battery charging issues. Like students in Alpha Middle School, students in HCPS were asked to bring their laptops to school fully charged, and the schools did provide ways to recharge the batteries during the day, but it would have been "helpful if charging were needed less frequently" (Zucker & McGhee, 2005, p. 24).

In addition to hardware, providing sufficient and appropriate software programs is another critical issue. The superintendent of HCPS, Dr. Mark Edwards, insightfully points out that "content is the fuel if a laptop is the car; it's film to the camera" (Zucker & McGhee, 2005, p. 5). The HCPS had a variety of software resources, from general applications to subject-specific software and tutorial applications to content created and disseminated by the school district. However, in many of the schools it was identified that more content-specific educational software was needed. Most schools have adequate generic productivity software but not enough educational software for specific subject matter content. For schools participating in the FTL program in Michigan, most of them had subject-specific software available on only about one quarter of their computers. Teachers expressed concerns about not having enough educational software. For example, a teacher mentioned: "I think that we have good hardware but can't get much of the good content-specific software that is available" (Urban-Lurain & Zhao, 2004).

Management

Computers are much more expensive than books and much easier to break. These fragile expensive devices bring new management problems, both at school and classroom level. At the school level, rarely has a school put items that cost about $1,000 into the hands and care of its students. Thus, there are serious policy issues that must be addressed; at the classroom level, teachers are facing new management challenges while trying to take advantage of the new devices.

School policy and regulations

In order to ensure that the laptops can be used effectively, many schools that have adopted one-to-one computing initiatives have adopted elaborate policies and regulations as well as provided tips to parents, students, and teachers on how to care for and use laptops. For example, the Enhanced Learning Strategy program in Canada has a 13-page iBook manual for students, a 19-page manual for parents, and a 58-page manual for teachers. The content in the manuals ranges from lists of responsibilities and requirements to tips on how to care for and handle laptops to instructions on how to use some of the laptop functions.[1]

In Alpha Middle School, a meeting was held for school administrators, teachers, and parents before the laptops were handed to students. Specific rules and regulations were discussed and made concerning policies on the safe use of the laptops, insurance, computer repair issues, and acceptable use, including regulations on privacy, Internet safety, and power management and rules on how to take care of laptops. Specifically, seven rules on having and using their laptops were clearly articulated to students:

"Seven" Rules for the Alpha Middle School Laptop Project

1. Keep your password secret.
2. You are responsible for keeping your laptop safe, secure, and undamaged. Take good care of it. Keep your laptop in its case when not used in class.
3. Bring your laptop to class charged and ready to go.
4. Use your laptop appropriately. No games, iTunes, iMovies, or other software on your laptop—unless it's for a class assignment. Visit only appropriate sites. Send only appropriate e-mail. Notify teachers of any problems or concerns.
5. Keep your laptop in a locked, secure place when it's not with you.
6. Understand that the laptop belongs to the school, and it can be checked at any time or taken away for disciplinary reasons.

7. The primary purpose of the laptop is for education. Treat it
as a valuable tool for learning.

Within this set of rules, three are about safety and security (1, 2, 5), three
are about the use of laptops (3, 4, 7), and one concerns discipline (6).

The laptop rules distributed to parents of students involved in the
Wireless Writing Program in Peace River North, Canada—such as
"Avoid eating and drinking near the laptop" and "Never leave the laptop unattended"—were all about how to take care of the laptops.[2] The
universal strong emphasis on the care and handling of laptops is under-
standable: Laptops are expensive and delicate digital pencils. They are
easy to break. Their portability—a designed advantage—makes them
more susceptible to theft than desktop computers are; hence, two major
concerns are laptop safety and security (Conway, 2005). Students were
given special training on how to take care of their laptops and keep
them safe and secure. One teacher required her students to treat their
laptops like a "newborn baby."

In some schools, even transportation was taken care of. For the safety
of laptops as well as of students, students had to use caution when they
got on and off school buses, and they were forbidden to play with their
computer backpacks on the bus. In one U.S. school district, a special
training workshop was held for school bus drivers so that students could
be better monitored on their way to and from school. An ICT consultant
in the United Kingdom reported that about 20% of the budget for the
Tablet PCs was spent on security measures such as storage facilities,
locks, screaming alarms, and engraving of the equipment prior to distribution (Twining et al., 2006). Due to concerns regarding students' safety,
some uses of the laptops were banned. For example, the e-mail capability
was blocked in HCPS because there were concerns about network integrity and online safety and security (Zucker & McGhee, 2005).

Safety and security are often the key pressures that influences
schools' chosen model of laptop deployment. An evaluation of the
early phase of a laptop initiative for students with literacy difficulties in
junior high school in Ireland documented how teachers in most schools
deployed laptops in a "fixed" fashion, utilizing them more like desktop
computers than mobile laptops (Conway, 2005; Daly & Conway, 2006).
In the United Kingdom, some schools do not allow students to take
their Tablet PCs home. But even in the schools that allow students to
take them home, many students choose not to (Twining et al., 2006).

Many unforeseen problems may arise during the implementation
of a laptop project, so old rules often have to be changed or new rules

have to be put in place. For example, although it was specified in the rules that students were not allowed to play games on laptops in Alpha Middle School, this activity still became one of the most popular uses of laptops soon after the project started. Some students played games in class. It gradually developed into a serious concern school-wide. To address this issue, a new rule was made: If a student was caught playing games in class, the teacher took his or her laptop to the principal's office, and then the student had to go down to the office and talk to the principal in order to take it back. This rule seemed to be effective because, as a teacher said, "I don't think we are seeing as much game playing as we used to because they know the expectations and they don't want to lose their computers." In one Maine school, to ensure that students brought their laptops to school, the principal made phone calls to the parents of any student who did not come to school with his or her computer and asked the parents to bring it to school (Zucker & McGhee, 2005).

Classroom management

The fact that every student has a laptop has also complicated classroom management. There are several issues that teachers need to deal with when using laptops in classrooms. One is technical problems. Computers can break down for all kinds of reasons; thus, very often too many students need help at the same time (Cuban, 1999), especially at the early stage of project implementation. According to the evaluation study on the FTL project, although teachers generally did not think that computers were difficult to use, one third of them reported that often too many students needed their help on computers, and it was difficult to get students to settle down after using computers in teaching (Urban-Lurain & Zhao, 2004) or control when to use and when not to use laptops (Rockman, 2003; Twining et al., 2006).

Another issue, which is more serious than technical problems, is discipline. Student discipline problems are not a new issue to teachers; nevertheless, it is a new challenge to monitor student laptop use in class. Take Internet access as an example. Teachers often found it distracting for students to have Internet access all the time. In class, students may check their e-mails, chat on instant messengers, play games, or just surf the Internet, and they are good at switching between programs when the teacher approaches. In our interviews in Alpha Middle School, a number of teachers expressed the wish for a special kind of software that could enable them to view all students' screens from the teacher station or to freeze a student's screen so that the teacher could check the programs opened on the screen.

Teachers also found it difficult to control when to use and when not to use laptops (Rockman et al., 2004; Twining et al., 2006). As a Virginia teacher noted, "Being the iBook 'police' is hard" (Zucker & McGhee, 2005). Teachers have to develop new classroom management strategies. An often mentioned strategy to deal with Internet distraction problems was to require students to close their computers when the teacher was talking. A teacher reported that she sometimes checks students' computers by pressing a certain function key that brings up all minimized programs.

The degree of the discipline problem in one-to-one computing classrooms varies greatly. The following two observations made in Alpha Middle School demonstrate the difference.

Classroom Observation 1

Five minutes after the bell rang, students in Mrs. A's language arts class were still trying to settle down. Mrs. A was making an announcement about an assignment she had asked students to work on, but the class was chaotic. Most students were busy on their computers without paying much attention to what she was talking about. One student raised his hand and said that he needed to go to the library to get his printout. Another student said that she did not have her computer so she had to go to the library to use the public computer. A third student said that his computer had run out of power so he had to go to the technical support staff to get a charger, and a fourth student had to run to another classroom because he forgot his paper there. Mrs. A was interrupted several times before finishing her announcement.

Classroom Observation 2

In Mr. B's social studies class, students were working on a WebQuest assignment on their own desks. They searched for the information they needed for their questions. One student had two windows open side by side: one was the question file in Word, and another was the Internet browser. She searched for the information she needed from the Internet and filled up her answer sheet in the Word document. Her teacher gave them a few links where they could find the information they needed. The teacher walked around the classroom, answering questions here and there. Students sometimes talked to each other about the links or the content. No discipline problem was observed.

One may argue that differences like these exist between classrooms without laptops. As a teacher points out, the discipline problem is an old problem. It just takes new forms with the laptops. Most of the classroom management strategies that worked in a traditional classroom also work in a ubiquitous computing environment.

Technical support

"Technology is great—when it works." Technologies can solve problems, but they themselves become problems when they do not work. Teachers and students inevitably encounter technical problems in their technology uses. If it is hard to get technical help or it takes too much time to solve the problems, they are less likely to use their laptops. Therefore, providing sufficient and timely technical support is very important to teacher and student technology use. A full-time technology coordinator may assist teachers with using computer software and hardware or adapting their teaching practice to include technology use.

Our interviews in Michigan schools and Alpha Middle School revealed that the availability of technology support not only influenced how much technology teachers used, but also affected which technologies they used. For example, the removal of Smart Board from classrooms in Alpha Middle School was, to a great extent, due to the fact that it broke down easily and timely technical support was not sufficiently provided. When the Smart Board was introduced to the school, teachers were excited about it because its functions were very impressive and it seemed to be very convenient. A teacher could sketch on the white board and the computer could record the sketching; the teacher could then e-mail the revised file to her students. However, after a few trials, teachers were frustrated because there were too many technical problems. It was very time consuming to set up and it seldom worked. In addition, a white board and a data projector (already equipped in their classrooms) serve almost the same purpose, but work far more easily and reliably. Therefore, the Smart Board was removed from this building (Lei, 2005).

Technical support provided by schools varies greatly in the different large one-to-one initiatives we examined. In the CN003 school in Hong Kong, the school technician helped teachers to carry out necessary technical tasks before the class (Hong Kong Case Study Report). In the HCPS in Virginia, every middle and high school had a full-time technology trainer who worked with teachers to integrate computers and other digital technologies into teaching and learning (Zucker & McGhee, 2005). However, in Maine, in the first year of the MLTI, about two thirds of the technology coordinators held another position in their school or district in addition to being the MLTI technology coordinator for assigned schools, and one third of these coordinators were responsible for providing services to more than one school (Silvernail & Harris, 2003).

Schools in the United Kingdom with Tablet PCs also reported "sharing" technicians or ICT consultants with other schools (Twining et al., 2006). Some of the schools relied on students for technical help (Silvernail & Lane, 2004; Twining et al., 2006). It was not surprising that many teachers reported that their school or district technology coordinators were not always accessible. Some teachers pointed out that the lack of technical support on a regular basis truly hindered the full use and potential of these computers in their schools, and too many technical problems could make it almost pointless to use the computers (Silvernail & Harris, 2003).

In the Michigan FTL program evaluation, it was found that, according to principals, about half of the technology problems could be fixed on the same day, nearly 20% of the problems needed up to 1 week to fix, and 8.2% of the problems could take more than 1 week to fix (Urban-Lurain & Zhao, 2004). Undoubtedly, the longer it takes to fix a technical problem, the more negative the effect that it has on teachers' and students' technology use.

Leadership

Successful implementation of educational technology requires coherent administrative support for using technology in education from the district level, through the principals to the teachers. According to the Stanford Research Institute (SRI) report, one major facilitator for the

HCPS project was that the "HCPS central office leadership was able to provide extensive support to rapidly expand and sustain the implementation of the laptops in this initiative" (Zucker & McGhee, 2005).

At the building level, principals are the key people who can directly influence the implementation of a laptop project. Principal leadership has been identified as one of the most important factors affecting the effective use of technology in classrooms (Daly, 2006; Daly & Conway, 2006). Daly identified what he termed a principal–patronage factor as a critical dimension of school culture in the implementation of the laptop initiative for students with dyslexia and other reading and writing difficulties (LISD) in Ireland. Supportive principals can not only model technology use but also highlight teachers' efforts to improve teaching and learning through technology use (Byrom, 1998). Energetic and committed school leaders can transform the entire school ecosystem into a technology-use-friendly environment in which teachers and students feel supported and encouraged to use technologies (Daly, 2006).

Supportive leadership is important in that it ensures established philosophy, clarified roles, and planned infrastructure to accommodate the efforts of any individual who wants to use technology. One lesson learned from HCPS's experience is that school leaders "need to articulate a broad vision and a systemic approach of how to introduce and use the laptops. Their breadth of vision contributed to the initiative's success and shows the importance of a comprehensive approach" (Zucker & McGhee, 2005, p. 27). In Hong Kong, a cutting-edge laptop project, "Cyber Art," that was started in 2000 was supported by an open-minded principal who not only secured resources but also provided enough flexibility and a suitable environment for this project (Hong Kong Case Study Report). On the other hand, the lack of supportive and strong leadership can directly lead to the failure of a laptop project implementation. Warschauer (2006) documented a case in which the departure of the principal, vice principal, and the laptop teacher at an elementary school resulted in transfer of the laptop project to another school.

At Alpha Middle School, the principal had a strong vision about technology use in this school. He set the tone; shared the vision with staff, teachers, and students; and enlisted their support. He gave teachers freedom to decide what technology they wanted to use and in what ways. He also recognized good practices of technology use and strongly advocated for these good practices. As mentioned by teachers during interviews, "He has created an environment that works out for everybody." Mrs. G, a language arts teacher, attributes the success of their laptop project mainly to the principal:

He is really our school leader. Without the leadership and the support we could never do this. Having a principal working towards our goals and being encouraging is really great. He has never faulted anyone for trying something new or taking a risk. So I think people saw that the support was there and the leadership was there. (Lei, 2005)

Principals need support from the school district to exert leadership. In most cases, the principals reported coherent support from their district's leadership. However, it is not uncommon that the school district may not support a school's laptop project, especially when only one or very few schools in the school district have laptop projects. Disapproval from the school district has proven to be detrimental or even disastrous to a laptop project.

Teachers

Teachers are a critical factor affecting laptop uses. As the first U.S. technology plan points out: "Indeed, without trained and experienced teachers, we know that computer equipment sits idle in classrooms, unused.... And we know that without high-quality software and well-trained teachers, computers alone do not help students meet challenging academic standards" (U.S. Department of Education, 1996). When teachers are not prepared to use new technology, computers end up being just souped-up typewriters (Cuban, 2001). Studies show that many teachers do not use the available technologies because they do not have the necessary skills and knowledge to use these technologies; some teachers readapt the technology according to their own technology skill level and their goals (Becker, 2001).

The importance of providing training to teachers has been well recognized. In the United Kingdom, professional development was provided to help teachers use the Tablet PCs. An ICT consultant pointed out, "Would you buy a £250,000 Bentley and never teach the person how to drive it properly? Well, the same goes for the Tablet PCs" (Twining et al., 2006), which echoes what an educator in the HCPS laptop project said about teacher technology preparation: "Teacher training was the most important reform effort in preparation for the iBook initiative" (Zucker & McGhee, 2005). In the FTL project, while principals, teachers, and technology coordinators would like to see more money spent on computers, they all agreed that the more important need was for training, both in terms of technology training and staff time spent on technology issues. All three groups agreed that simply providing more computers would not be sufficient (Urban-Lurain & Zhao, 2004).

Some states have specific requirements on teachers' technology proficiency. For instance, Virginia requires teachers to present a portfolio showing mastery of certain technology skills. The HCPS also requires new teachers to complete 12 hours of technology training (Zucker & McGhee, 2005). Some one-to-one computing schools had a series of strategies for preparing teachers to use laptops in their classrooms, such as encouraging teachers to take college educational technology courses, providing local professional development workshops, pairing less experienced teachers with advanced teachers, and giving teachers time and encouraging them to play with computers (Lei, 2005; Urban-Lurain & Zhao, 2004).

Providing formal professional development to teachers

Convenient, effective, and ample professional development opportunities are important in preparing teachers to use technology in a one-to-one computing environment. In Alpha Middle School, teachers were encouraged to take master-level courses on educational technology. The school (school district) provided teachers with many professional development opportunities, strongly encouraged teachers to learn, and was able to "make it very easy for teachers to take technology classes" (Lei, 2005). Most of the training was on-site training and the school was very open in giving teachers the support they needed to work with technology. The technology specialist and a number of technologically savvy teachers gave workshops on how to use specific applications such as Word and PowerPoint, as well as Web page design and Internet sites for education.

These workshops were more useful for teachers than those from outside because they were tailored to teachers' local needs and were more usable in classrooms. This school also had regular teacher team meetings. Teachers teaching the same grade met at least twice a week, one period each time. The technology specialist often came to team meetings, introduced new technology or education resources, and explained how to use them. Team meetings also provided a venue for teachers to share their experiences and ideas on using technology (Lei, 2005).

In HCPS, in addition to having a full-time technology trainer in every school, workshops and classes about technology integration were available to teachers during the academic year and in the summer (Zucker & McGhee, 2005). In the LISD in Ireland, a number of all-day meetings were held for teachers and principals. Workshop sessions for teachers focused on identifying appropriate software for use with students experiencing literacy difficulties (Conway, 2005; Daly, 2006)

Incentives were often used to motivate teachers to learn how to use technology. In Hong Kong, for example, teachers received discounts when taking technology preparation courses. Teachers in Alpha Middle School received college credits for attending workshops. Teachers in HCPS were paid an $18/hour stipend to take classes ranging in length from a day or less to as long as 2 weeks (Zucker & McGhee, 2005).

Providing time for teachers to explore

The lack of time for teachers to learn, practice, or plan ways to use computers has been identified as one barrier for teacher technology use (U.S. Department of Education, 2000b; Zhao, Frank, & Ellefson, in press). A supportive context should provide teachers time to explore and get acquainted with technology. Experimenting with technologies can help teachers and students in many ways. Papert (1992) suggests that any kind of "playing with problems" enhances the abilities that lie behind their solution because spending relaxed time with a problem leads to getting to know it and, through this, to improving one's ability to deal with similar problems (p. 87). Burbules and Callister (2000) also point out that just "messing around" is an indispensable approach for users at all levels of sophistication because, in doing so, they not only have a chance to work their way out of the problem, but also have an opportunity to discover new capabilities of the system they are using.

In the United Kingdom, the deployment of the Tablet PCs was very different across the schools. In some schools, the Tablet PCs were shared, and in some schools they were owned on a one-to-one basis. Researchers have found that ownership was a determinant of usage: Teachers who owned the Tablet PCs reported much more extensive usage because individual ownership provided the needed time to experiment with the PCs to establish confidence in incorporating them into teaching (Twining et al., 2006). Similarly, Zhao and his colleagues (in press) find that teachers reported more computer usage when they had opportunities to explore technologies on their own, and thus they suggest that schools make the effort to build in "play" time for teachers during the school day and professional development sessions to allow teachers to explore technologies on their own. One teacher in Alpha Middle School talked about his positive experience playing with computers:

I think the play mode is the best way to learn and that is how I learned with iMovie. I didn't know how to use it but I wanted to do a movie project ... so I just played with it and learned how to manipulate the tools and now I know how to do it, and eventually I was able to teach my students how to use this. So I think the play—just sitting down and having time to experiment—is one of the best ways to learn how to do it.

Playing with technology also helps change teachers' attitudes to and confidence in using technology. For example, one of the teachers in LISD of Ireland had little or no experience with technology use prior to the laptop initiative, but over a 6-month period her change was surprising to the researchers:

Her confidence in using and thinking about the possibilities of using laptops increased over a few months from our first to our second visit. While nothing concrete or new has occurred in this school in recent months, the horizons of possibility have expanded and a new confidence and relaxed attitude has emerged in the school's perception of technology. The learning support teacher admits to being nervous of technology but has learned to "play" with it. (Conway, 2005, p. 151)

However, it is often the case that teachers do not have time to explore technology. According to the National Report on NetDay's 2005 Speak Out Event, 57% of teachers report that the number one obstacle they face in using technology at school for professional tasks is lack of time during the school day (NetDay, 2005). In schools in Michigan, teachers complained, "We get a lot of money to purchase technology equipment and software, but we are given no time to learn how to use it unless we do it on our own time and receive a small stipend" (Urban-Lurain & Zhao, 2004, p. 48).

Similar complaints were heard in other projects (Conway, 2005). For instance, the phase one evaluation on Maine's one-to-one computing initiative revealed that the lack of sufficient time for teachers to become more skilled technically in using the laptops and more skilled pedagogically in integrating the laptops into their instruction was a serious concern. Teachers repeatedly expressed the need to have release time to explore, learn, and study computer software (Silvernail

& Lane, 2004, p. 15). Teachers in HCPS were more fortunate in this regard. The school district paid many of its teachers to help develop lessons and other resources for using the iBook computers (Zucker & McGhee, 2005).

Teaming up teachers

Even if sufficient time is provided, sometimes it is still difficult for teachers to learn about technology on their own because, as a teacher said, "Sometimes when you are not computer savvy and you don't know how to use it, then it is difficult." Moreover, not everyone likes to play with computers. Many teachers also view spending time on exploring how to use technology as a waste of time, especially when they have assignments to grade, lectures to prepare, and parents to call. They would rather ask somebody who knows how to use a specific technology instead of exploring on their own because why "reinvent the wheel"?

In every school, there are always teachers who are very good at using technology and those who do not have much experience in teaching with it. It often works well if these two groups of teachers are teamed up. One advanced teacher can lead one or more teachers who do not know much about technology. The CN003 school in Hong Kong established a "satellite group" to encourage teachers with higher computer proficiency to provide guidance for those who were unfamiliar with using technology for teaching (Hong Kong Case Study Report). During an interview at Alpha Middle School, one technologically savvy teacher mentioned how he helped his colleagues learn the Excel program:

> I know how to use Excel to set up a grade book in Excel and calculate grades using formulas. There are a lot of teachers who don't use it because they don't know how. One day I sat down with a couple of teachers and showed them how to use it and now that is what they use.

Learning from a colleague also makes technology look much easier because teachers can learn from real examples on how to use specific technology for specific tasks.

Informal help from colleagues was ranked as the most effective professional development activity by teachers in Maine. Many teachers indicated that they had learned a great deal about technology from colleagues (94%) and on their own (93%) (Silvernail & Harris, 2003).

Sometimes, it is a matter of knowledge exchange: You know something I don't know, and I know something you don't. When we work together we can learn from each other and explore more.

Making it easier for teachers

Some technologies are quite complicated and time consuming, such as computer programming or multimedia production. It is generally neither easy nor necessary for all teachers to learn how to use them. But sometimes teachers do want to utilize these more sophisticated applications in their teaching. It would be regrettable if they were deterred by the complexity of these applications.

One way to solve this problem is to simplify the use of complicated applications. For example, in Alpha Middle School, only a few teachers knew how to create and design Web pages. In the past only these few people had their own personalized course Websites. Other teachers used the general course management system. Gradually, more teachers wanted to have their own course Web site that could be more flexible and personalized. Thus, Mr. Bob, the most technologically savvy teacher in this school besides the technical support team, developed several Web page templates, e-mailed them to other teachers, and spent some time at regular teacher team meetings to explain how to use these templates to meet individual needs. Similarly, game templates based on PowerPoint or Excel, such as the Who Wants to Be a Millionaire? game or the Jeopardy game, can be provided to teachers so that they can easily adapt these popular games in their subject areas.

In this case, teachers do not need to understand the concepts and fundamental operations behind the computer screen for a specific program. Instead of learning programming language and Web page design, they just need to apply knowledge and skills that they already know to a new setting. This greatly reduces both the difficulty of the task and the anxiety level of using advanced technology.

Providing help during transition from traditional classrooms to ubiquitous computing classrooms

For a laptop project to succeed, it is very important to help the teachers to make the transition from a traditional school environment to a ubiquitous computing environment. This transition, for many teachers, is very challenging, especially for those who do not have much technology experience, do not like to take risks, and have been working in a traditional school environment for years. When talking about the laptop initiative, a principal in HCPS said, "If it's just dropped on teachers, it's a problem" (Zucker & McGhee, 2005). If the technologies are imposed upon them, teachers might be overwhelmed by the changes and challenges and thus cannot see the opportunities created by laptops. Their reactions to laptop projects would be just as one teacher pointed out during an interview in Alpha Middle School:

> A: they are going to be resistant because they don't know how to use the laptops; and B: they will just be overwhelmed with what they have to do. I think you have to transit into it gradually so that teachers have a chance to adjust and adapt to what is coming.

This gradual transition can be done in a number of ways. First, as teachers in Maine pointed out: "Teachers should have had laptops much earlier in order to prepare for the upcoming year" (Silvernail & Harris, 2003). Teachers need time to warm up for the one-to-one initiative. As one teacher in the LISD in Ireland noted, the first few months after receiving the laptop constituted a "scratching our head phase." That is, the teachers need time to figure out how to use the laptops themselves, how they might play a role in the curriculum, what software might be productive for students with literacy difficulties, and how best to deploy laptops within the existing provision of learning support for students with different needs (Conway, 2005).

Second, some schools had mobile laptop carts before they launched a one-to-one laptop project (e.g., Lei & Zhao, 2005; Zucker & McGhee, 2005). They can experiment with these carts and explore how to use laptops when students all have their own laptops—not on a daily basis, but for some classes. This can help teachers build knowledge and experience of how to use the technology, and reduce anxiety in a ubiquitous computing environment. They become familiar with using technology in their classroom, so the transition from a traditional classroom to a one-to-one classroom can be smoother.

Another way is to prepare teachers through professional development. Teachers and staff can be trained in steps to become comfortable gradually with the laptops and with the idea of one-to-one computing. When they feel confident, they can see the possibilities and can embrace them.

The HCPS experience serves as a good example. According to the SRI report conducted by Zucker and McGhee (2005), the laptop project in the middle schools in HCPS took more than a year to prepare in several steps. First, around December 2001 or January 2002, middle school teachers received their iBooks, so they had a full year to use them before their students received laptops. In fall 2002, middle school students received computer training from the physical education department; at that time, the iBooks were on portable carts. The PE teachers created lessons for students that focused on the proper use of the computers. Middle school parents were also provided many training opportunities during fall 2002. Finally, at the end of January 2003, the middle school students received their laptops. This step-by-step approach provided ample time for teachers to adapt gradually to a ubiquitous computing environment.

Developing a supportive school environment/culture

A school environment that provides rich social capital and peer support, supportive social networks, and an encouraging culture is also important in promoting technology uses. It encourages teachers and students to explore new technology, experiment with new teaching practices, work collaboratively, and learn from each other.

Social capital and peer support

Schools are dynamic, human, and intensely social places. In this social environment, people are related to each other and have mutual influences on each other's decisions about what and how technology is used. A student in Alpha Middle School said during his interview, "Everyone influences each other, because everyone has his own way." This mutual influence is reflected in how students learn technology, as another student commented, "Sometimes we learn from our teachers, sometimes from friends. We tell each other what we've learned and found out. That's how we learn." When talking about how she learned the Photoshop program, Alison, an eighth grader from the same school, said, "It's really kind of like 'monkey see, monkey do.' My good friend was working on her picture on Photoshop, and I said, 'Wow, can you show me how you do that?'"

Similarly, teachers can serve as resources to each other. If teachers share their ideas or technical skills with their peers, the knowledge can be disseminated more quickly. A technology coordinator in HCPS said:

> Having teachers as resources is a key.... We have a ton of this spirit at this school—teachers reaching out for help to other teachers. A lot of it is framing it to teachers to show them the benefits of using technology. Professional development has to be ongoing at your site. We encourage teachers to "steal" from each other. (Zucker & McGhee, 2005, p. 23)

According to the National Report on NetDay's 2005 Speak Out Event, 13% of teachers surveyed reported learning new technology from professional development opportunities, while 23% of them reported learning new technology from peers (NetDay, 2005)

Social connectedness

How socially connected the teachers are also significantly influences teachers' technology use. This can be explained from a number of angles. First, more socially connected teachers can get more technology support from their colleagues. This informal technology support significantly affects how much technology they use. Second, more socially connected teachers have more channels to get information related to technology use, such as subject-related resources, tips on using specific technologies, and professional development opportunities and choices. This type of information can help teachers make well-informed decisions. Third, social connectedness not only influences how users learn a new technology, but also affects how they use technology by providing related information and exerting pressure to use the technology (Lei, 2005). Therefore, more socially connected teachers are under higher peer pressure related to technology use, especially in an environment where most teachers are trying hard to incorporate technology into their teaching.

More socially connected teachers may also be able to get technical help from their friends in addition to school technology specialists. For instance, when asked where she could get help when something did not work, Julie, an eighth grade student, said that she got help from her friends: "My friends, I think they know more about it than I do, so when I don't know how to do things, they usually help me out" (Lei, 2005).

School culture

The "culture of the building" is one of the critical issues for successful technology integration identified by research on one-to-one computing projects (Bonifaz & Zucker, 2004; Daly, 2006). The school culture influences people's value judgment, decision-making, and social behavior. In a school where risk-taking is viewed as stupid or trying something new is equated with triviality, technology innovation is not likely to happen. On the other hand, if innovation is recognized and valued, more people will try different ways of using technology. In a school where "there are very few people who are not willing to try new things," a laptop project is more likely to succeed (Lei, 2005).

A supportive school culture nurtures a healthy community. It was found that teachers who shared information with colleagues, collaborated on instructional activities, and maintained a reflective professional dialogue used complex technologies and integrated them into teaching more effectively than teachers who were not part of a strong professional community. A technical specialist in Virginia points out that "relationships and trust are very important to the success of the initiative" (Zucker & McGhee, 2005, p. 19).

Also important is the relationship and trust between the school and the local community. As high-investment and high-stake initiatives, one-to-one computing projects often arouse much controversy or even confrontation, especially at the early stage. In the first year of the HCPS one-to-one laptop project, the schools faced a number of challenges from the community, such as parental concerns about inappropriate student use of the computers and some negative articles about the laptop initiative published by the local press. This situation gradually changed as communication was improved and a sense of community was fostered (Zucker & McGhee, 2005).

If we borrow the famous sentence of Tolstoy's great novel *Anna Karenina,* "Happy families are all alike; every unhappy family is unhappy in its own way," we can also say that "successful one-to-one projects are all alike; every unsuccessful project is unsuccessful in its own way." By this we mean that a successful one-to-one computing project has to succeed in many respects: budget, technology, resources, teacher training, school policy, management, leadership, community, school culture, etc. If any of these conditions is not provided, it is very difficult for a laptop project to survive, let alone to succeed.

In some cases, because of many management and infrastructure problems, laptops are locked up and turned into desktops. Mobile technology loses its mobility and becomes immobile, and the laptop initiatives lose their early intention of learning "anytime, anywhere."

These factors' importance changes over time

In the previous section we discussed the conditions that influence laptop use. These conditions are not independent of one another; their influence on laptop use varies from case to case, and their interactions and relationships change as the school environment evolves with the implementation of the laptop project. This is because the use of laptops in schools is not independent or a series of isolated events; rather, it is situated in and connected with the school context and is an integral part of an ongoing process of change.

The school context is gradually evolving after the introduction of laptops, changing the beliefs, behaviors, and abilities of teachers and students, which further changes the challenges the school faces at different stages. This, consequently, requires changes in the conditions to facilitate the continuing implementation of laptop projects. Based on our study in the Alpha Middle School over 3 years, Table 3.1 summarizes the characteristics of different factors at the early and later stages of the implementation of one-to-one computing projects.

As Table 3.1 suggests, at the beginning of a one-to-one laptop project, technological proficiency of the teachers and students varies a great deal, depending on their previous experiences and training in technology use. But, generally speaking, they do not have much technological knowledge and skill, and most of them need training about laptop use in general and software applications in particular. Most teachers and students have high expectations of and high motivation for using laptops. This excitement is clearly reflected in surveys as well as interviews at the earlier stage of the laptop project in Alpha Middle School. For example, laptops were believed to be able to "increase learning, encourage communication, motivate students" and help students to "hand in homework on time easily, having classes online, no assignment will be late or lost." Other comments included: "I think the laptops will improve education; they can make wrong answers into correct ones" and "Everything will be easier." The use of laptops at this stage focuses on general applications such as using Word to take notes, e-mailing, and surfing online. The major focus of project implementation is to put these laptops in use.

Table 3.1 Changes in school context, laptop use, and conditions for laptop use over time

School context	Characteristics of school context at different stages of laptop implementation	
	Early stage	Later stage
Teachers and students	Technology proficiency varies greatly; most people need training in technology use; high motivation and high expectations; having laptops is viewed as a privilege	Sufficient knowledge and skills in using laptops; some people are very advanced in technology; having laptops is taken for granted
Laptop use	Mostly basic operations using general applications	Much more diversified and for various purposes; more use of subject-specific applications and multimedia applications in addition to general applications
School culture	Focusing on increasing the quantity of use	More attention shifts to the quality of use; concerns emphasize misuse and abuse of laptops
Conditions	Focus at early stage	Focus at later stage
Technology infrastructure	Laptops are brand new, functioning well; providing supporting technology hardware and software is a major issue	Laptops wearing down, more investment is needed to upgrade software and replace broken/lost hardware
Management	Policy and regulations on safety, security, and caring for laptops	Policy and regulations on appropriate use of laptops
Technical support	Fixing problems caused by unfamiliarity with laptop use and low technology proficiency	Fixing problems caused by hardware wearing down and breaking down
Professional development	To increase technology proficiency and enable the general use of laptops; providing resources for using technology	Helping teachers integrate technology in teaching in more meaningful ways; providing resources in subject areas

At the early stage, the laptops are brand new and are generally functioning well. If enough laptops are provided, in terms of technology infrastructure, supporting equipment such as CD burners, projectors, printers, and software is needed. Since laptops are new to most students and they do not know how to care for and use their laptops, management issues mainly concern safety and security policies. Students need to follow these regulations to take care of their laptops and avoid losing or breaking them. Technical support is in great demand because most people do not know how to use their laptops.

Much confusion and many problems are caused by users' unfamiliarity with the laptop. For example, taking notes on laptops is the most common use immediately after students receive their laptops. However, for many students who do not have much technology experience, it is not easy to remember to save their working documents in particular folders and they have problems retrieving the files when they are needed. Many do not know how to back up files, so once anything happens to their laptops, there is a big panic. Therefore, it is critical to help both teachers and students learn basic computer operations, increase their technology proficiency, and enable the general use of laptops. Resources on how to use technology are also important so that they can learn on their own.

As the laptop project progresses, teachers and students gradually develop increasing expertise in using technology. They do not have as many problems in using general applications as they did at the beginning of the project. The uses they put their laptops to are becoming increasingly diversified: from basic uses such as typing, e-mailing, and writing to playing games, creating PowerPoint presentations, and using subject-specific software for more advanced and complex uses, such as creating Websites, composing music, and making movies. The purposes for which they use their laptops increase in diversity: from learning and communication to self-expression, entertainment, or multimedia production. Their attitudes toward having their own laptops also change as time passes and they get used to the idea. Once a privilege, now laptops become a natural part of their lives. When talking about the difference between students now and 2 years ago when the laptop project had just launched, a teacher in Alpha Middle School said:

> When we first gave the computers out, people were in awe of them. Now it is kind of a given. In other words, they have had more exposure to it now and it is certainly more of a social occasion tool … It is almost as if it is the expected.

At this stage, the school culture regarding laptop use gradually shifts from emphasizing the quantity of use to the quality of use. Instead of focusing on whether or not laptops are being used enough, more attention is being paid to how they are being used. The questions now asked are how to use laptops in meaningful ways to help students learn better, how to regulate students' Internet use, and how to avoid misuse or abuse of laptops. In a dynamic system, there are always new issues that need to be addressed; thus, new regulation and rules may need to be put in place. For example, in Alpha Middle School, teachers were not given administrator access codes to the laptops at the beginning of the project because most teachers were not technologically savvy. The technology department was concerned that teachers would do damage to the laptops. As teachers built up expertise in using technology, trust between teachers and technical support staff also gradually built up. Then, in the second year, all teachers were given administrator access. For students, new regulations are often related to misuse of technology. For instance, when teachers and parents found that students were spending too much time chatting with friends on instant messengers, chatting in school was soon banned.

One might expect that, as teachers and students become more savvy technologically, there will not be as much need for technical support as at the beginning of the laptop project. This, however, is not what we found from our studies. Technical support is always a critical condition for successful implementation of one-to-one computing projects, and evidence shows that there is an increasing need for technical support. At least two factors contribute to the increased demand: the technical problems caused by natural wearing down of technologies and the technological assistance needed by teachers and students as they explore more advanced technology applications.

Information technology is advancing at an incredibly rapid rate. After 2 or 3 years, laptops become outdated and may not be able to perform certain tasks and support certain applications. Hardware wearing down further exacerbates the problem. During an interview in the third year of the laptop project in Alpha Middle school, the technical specialist reported:

> We are getting some issues with the machines that we didn't get in the beginning, such as the normal wearing down of batteries. This is something we never had to worry about. I don't want to say the laptops have a delicate design but with the kids opening and closing them seven or eight periods a day and taking them home, I just question the sturdiness.

After having laptops for 2 years, over 90% of students in Alpha Middle School reported that laptop breakdown was the biggest problem and most annoying aspect of the laptop project. A student complained in a survey: "My laptop is probably one of the worst quality laptops. It is sometimes slow and crashes all the time; all of my friends have had their computers crashed or broken already."

The need for technical support is growing; meanwhile, the urgency of fixing a technical problem is also increasing because students are increasingly used to having their own laptops and doing all their work on their laptops. Laptops have become an integral part of student life, so some students just do not do anything if their laptops are under repair. Some teachers complain that students have forgotten that they can still learn with paper and pencil. Not having their laptops makes them nervous and "they feel they need it right away."

Teachers also constantly need technological assistance. As they become more skilled and experienced in using them for teaching, the technical assistance they need also becomes more advanced. From learning basic knowledge and skills, to trying to integrate technology in their teaching, to looking for new applications for their subjects, teachers' need for technological assistance gradually changes from basic technology use to more meaningful technology integration. This shift is reflected in the teacher professional development provided by the school.

Summary

Similarly to the diverse nature of natural ecosystems, schools across the world are unique in their own ways. A third grade classroom in an American school is different from a third grade classroom in a Chinese school in many ways, such as how the classroom is arranged, how classes are scheduled and conducted, what knowledge and skills are taught, and what technologies are used and in what ways. Similarly, a Japanese classroom is very different from a Vietnamese or Thai classroom. The dynamics in different schools are also different. These differences pose different challenges to the implementation of one-to-one laptop initiatives.

However, despite the differences, it is still possible to find commonalities in the enabling conditions for one-to-one laptop use in schools. In this chapter we summarized the common themes reflected in laptop studies in several countries and regions. It is clear that it takes much more than the idea to make it work. Technology integration in schools is never an easy task. The successful implementation of a ubiquitous computing project depends not only on a working technological infrastructure, which ensures that the technology can be used, but also, more importantly, on an effective human infrastructure that supports and facilitates the meaningful use of technology. Even in a school where a large, ubiquitous computing project has been successful, to sustain the success requires adjustment in policies, continuous financial investment, increased support, and consistently strong leadership.

How students use laptop computers

[W]e cannot specify the pure, or ideal, case for the *use* of an innovation, only its idealization in the minds of the developers. Users inevitably interpret an innovation in distinctive ways, apply it idiosyncratically in their own contexts, and even re-create it to satisfy their own needs.

Bertram (Chip) Bruce, *Innovation and Social Change*, 1993

Introduction

What do students do with their laptops? What happens when each and every student in a class has a laptop all the time and the laptop is networked? We know that adults seem to have a picture in mind of how the laptops should be used. The ideal picture is that students use the laptops primarily, if not exclusively, for learning activities. This is clearly stated in the "Alpha Middle School Seven Rules for the Laptop Project":

> 7. The primary purpose of the laptop is for *education*. Treat it as a valuable tool for learning.

In this context, the word "education" can be understood as "learning" or "subject learning". In contrast to "learning activities," activities that are not directly related to specific subjects, such as playing games and online chatting, are generally not encouraged or are even prohibited, as stated in the same document:

> 4. Use your laptop appropriately. No games, iTunes, iMovies, or other software on your laptop—unless it's for a class assignment. Visit only appropriate sites. Send only appropriate email. Notify teachers of any problems or concerns.

These two rules indicate that adults want to separate learning from playing, and that laptops are only intended for learning, not for playing, whether in or out of school. But do children follow these rules and use laptops in ways prescribed or expected by adults? In this chapter, we discuss the various uses students have been found to make of their laptops in the many one-to-one computing initiatives.

Laptop for learning

> I can search and find a lot of information fast and I can contact my teachers and friends about my school work or about having friends over. The laptops made my life easier because we could type, research and take notes at the same time.
>
> The laptops are good for taking notes, watching movies for science, interacting, learning, researching online and doing WebQuest.
>
> **Students in Alpha Middle School**

Laptop projects in schools worldwide were all set up for learning purposes, and research found that laptop use in schools indeed was mostly for learning purposes. From physical education to foreign languages, laptops were used in all subjects, with the heaviest uses in English and language arts, science, and social studies. Some uses of the laptops were common across many subjects—for instance, taking notes with word processing programs, searching information online, and accessing online libraries that provide materials for all subjects, such as unitedstreaming video library and the Library of Congress. Some were specific to a particular subject, such as the ALEKS program for math or desktop publishing for language arts. Here we look at some of the most common laptop uses for learning.

Taking notes

Taking notes is one of the most common uses of the laptops in all the schools with one-to-one laptop projects. Students use word processing software such as Microsoft Word, Word Perfect, and Notepad to take notes in class. Some students even take notes in PowerPoint. Instead of carrying a pile of notebooks for different subjects, they now carry their laptop "notebook" from one classroom to the next. Sometimes teachers send their PowerPoint presentations to students before class, so during the class when the teacher is explaining a concept from the PowerPoint presentation, students have the same PowerPoint file open on their laptops and take notes in the note section on the same slides.

The portability of laptops makes it easy to take them on field trips. Sixth grade students in St. Mary's RC Primary School in the United Kingdom took their Tablet PCs with them on a field trip to the Isle of Wight. They took notes on their Tablet PCs and used the audio recording program to record their diaries every night. After the trip they compiled a multimedia presentation for their parents (Twining et al., 2006). The students who participated in the Cyber Art Project in Hong Kong took their laptops with them on a trip to Beijing and took notes and drew pictures of the Great Wall and the Forbidden City (Hong Kong Case Report). For the science class, students took their laptops to fields to record data and take notes on their observations.

Students liked their new "notebook." The easy editing and spell-checking features of word processing programs helped students keep neat notes with fewer grammar and spelling errors. Students in the United Kingdom who had Tablet PCs used art and drawing facilities when taking notes (Twining et al., 2006). They also learned how to organize their notes for different subjects so that they could easily find them when needed. The folders were "like a binder," but more organized. Students enjoyed exploring and exchanging little tricks on how to better organize their folders. For example, instead of using the briefcase icon for the folders, they figured out that there were many different icons they could choose for different folders and they could also put a picture on a folder as a reminder of the content. Therefore, they could personalize their folder trees in ways they liked.

The personalized folder trees help students themselves to be more organized. A student in the Alpha Middle School told us:

> I used to keep a lot of things in my binder. I am not very well organized. I am messy. That's my problem. In the middle of the year, I started taking notes on the computer, organizing folders. I don't lose my notes any more. I am better prepared for my tasks [and] tests.

Another student viewed himself as a well-organized person, but believed that "it helps those unorganized kids a lot." Students in other laptop projects in the country also described learning note-taking skills via iBooks when some of their teachers provided "skeleton" notes and required students to fill in the blanks as the lesson progressed (Zucker & McGhee, 2005).

In Alpha Middle School, using "stickies" to take informal notes was very popular. Stickies are desktop notes, or the virtual Post-It, serving as reminders of important things to do. Once on the screen, they remain where they were placed until they are closed. Students use these stickies to keep notes of their homework, what to do during study halls, questions they need to ask the teacher or a friend, and what to take home after class. They were so popular that they even replaced the assignment book:

> In the beginning of the year, we all got an assignment book. We were supposed to keep our assignment book and write down all our assignments, but now we have the stickies notes on our computers; it's like a poster note, but it's on our computer. So now we use these stickies notes instead of our assignment book.

Typing notes was also preferred to handwriting by most students, especially those who did not like handwriting or did not think their handwriting was good. A student said, "My hand doesn't cramp from writing and all my work is in one place." Another student liked taking notes on his laptop because it was "better for taking quick neat notes because I have sloppy handwriting and handwrite slowly." In addition, they can also share and compare their notes. If a student was absent from a class, she could get electronic copies from her classmates. In some cases, students' notes are projected and shared with the whole class to reach a class consensus (Twinings et al., 2006)

Searching for information on the Internet

With more diverse, complete, and updated information than textbooks, the Internet opens a vast information resource for students. Immediate access to the Internet allows more "just in time" learning or incidental learning (Lei & Zhao, 2006; Warschauer, 2006) Searching out information on the Internet is naturally one of the most popular and frequent uses of laptops for learning purposes. Students search online for different subjects for various tasks, such as finding additional information, learning specific subject content, and working on course projects.

A student remarked, "They [laptops] contribute to my education by allowing me to surf the Internet and learn about the world and current events, not just the information we learn in dated books." That is what students do. They often research on the Internet to find more information on what is being learned in addition to information provided in the textbook. For example, in a language arts classroom, students were studying *Racing the Sun* by Paul Pitts. After reading the text, they researched the author's biography online, learned about his other works, and finally created their own cast for the story. Similarly, to help bilingual students better understand the epic poem *Beowulf*, students did research online about the background of this poem, such as lives in ancient times; read multimedia narratives on parts of this story; and then designed their own modern newspaper on Beowulf's journey and battles (Warschauer, 2006).

In a history class, the teacher provided a list of Web sites where students could find the most current events happening in the state, the country, and the world. During the U.S. presidential election in 2004, students in Alpha Middle School tracked the state primaries online to find information about the candidates, follow news on who was running, and check the results of the elections at different levels. A social studies teacher in California commented, "I can make the Zulu tribe in South Africa come alive for them, versus talk about these people that they've never seen, they don't know, they can't grasp" (Warschauer, 2006).

Instead of going to the library searching in the books, students are more likely to open their Internet browsers and search for information online. In schools with one-to-one computing, a notable change is that student visits to the library dramatically decreased soon after students had their own laptops, because they turned to the Internet for information. However, the frequency of library visits may go up again, not for borrowing books, but for asking help from the librarian on how to search for credible information online. This generally requires a change in the role of the librarian, from a traditional librarian who helps students find, check in, and check out books to a media center librarian who helps students search for information, identify online resources for the school, and sometimes provide information literacy education to students (Lei, 2005).

For students, the best thing about having a laptop is that "you can go to so many places around the world without getting out of your desk and get all the information you want," and this makes them feel powerful because of "having all the information in the world at my fingertips."

Learning specific subject content

Students also go online to learn about specific subject content. Many subject matter teachers have their favorite Web sites or online programs where students can learn new content, complete exercises, take a quiz, work on projects, and discuss and debate with their classmates. Online learning is extensively used for science and math. A number of Web sites and online programs are being used for math class, such as Geometer's Sketchpad, Carnegie Algebra, and ALEKS. The Carnegie Algebra program provides the procedure for solving a problem, but students have to figure out the logic of how to solve the problem. The program can display the student's work results at each step, give warnings when the logic is wrong, and provide hints for solving a problem. The Geometer's Sketchpad is an interactive software program for learning geometry where students can construct an object and explore its geometrical properties by manipulating different parts of the object with the mouse.

Programs like these provide students with a "virtual playground" (Bartels, 2000), where they can construct, explore, and learn at their own pace. A teacher in Alpha Middle School mentioned his favorite site, National Library of Virtual Manipulatives, which he introduced to his students: "In my class, we go to one specific site called National Library of Virtual Manipulatives. Anything we played in school like the cube, square … anything you can manipulate with your hands, they have it online virtually."

In science classes, students often go to specific Web sites such as the NASA Web site to collect materials, observe, analyze data, and then summarize what they have learned. For example, when learning about moon cycles, students went online, studied moon faces on a calendar, and figured out the cycles. When learning about continental drift, they may go to a Web site that presents information about theories and models of continental drift or look at a diagram or a simulation, in order to arrive at a more concrete understanding of this concept. For astronomy, students go to the NASA Web site, study the pictures of different planets, and learn about the major planets and their relationships.

Laptops have been used for many purposes in science class, including "cultivating the skills necessary for scientific inquiry: generating research questions; formulating hypotheses or predictions; developing models to describe or explain a phenomenon; and collecting, displaying, and analyzing data" (Zucker & McGhee, 2005, p. 12). In Virginia, the researchers made the following observation:

> The goal of one observed physics lesson was to review what students had learned about how to add and subtract colors. The lesson included an exercise in which students used the laptops to create color images that contained the primary colors. Several other activities were done

by using a Web site tutorial that allows students to create different colors by manipulating overlapping ovals of complementary colors. While all the students worked in pairs on their laptops to manipulate the color combinations, the teacher moved through the class, helping students by posing questions and providing clarifying information. Students talked with each other about the activity and offered assistance to other groups of students. (p. 14)

The availability of a large number of online simulation programs has in essence made the online environment a virtual science lab for students. In these virtual labs students can easily manipulate and control natural forces and processes. This allows for easy, quick, and safe experiments with different designs. An eighth grade science teacher reported:

I am looking for simulation, an online lab where kids can go, they can manipulate things on their own computer and see how things react. For Newton's law, they have an online acceleration track; you can change the mass or the friction and run it, and it gives you data. There is also an online virtual earthquake where you can figure out why the earthquake happens.

Students also use laptops and Internet resources to take tests. In Henrico County Public Schools, students had access to SAT tutorial software and to online practice for state tests, took customized online tests and quizzes through Quia, and found resources aligned to state standards through a special search engine, netTrekker (Zucker & McGhee, 2005).

Online discussion

Online discussion boards provide students the capacity to ask questions, voice their opinions, exchange ideas, and sometimes critique and be critiqued. In Alpha Middle School, all seventh grade students visit their social science class Web site regularly and post their messages as required by the teacher:

I've had a lot [of] online discussions. The kids overwhelmingly respond to that positively. I ask them to post the questions in the class, and then they have to post twice more within the next three days. The quality of some of them is extremely high, some are seventh grade kids. The quantity of their posting is so much different than if I just give them a piece of paper and say write them. Clearly they are more engaged in what they are doing.

In teacher Howard's math class, students posted their proofs, including components such as the construction, figures, and steps of how they arrived at their proof, on a Web discussion board. They also commented on each other's proofs and suggested how they could be more efficient. They received points for logging on to the discussion board and posting messages. The students' discussions on this board were quite meaningful and deep. As the teacher commented, "They critique each other … they say 'I look at so-and-so's proof and I think I can improve it,' or 'Do we have to say this? I think it's redundant.'" Online discussion like this gives students an opportunity to think more about what they learn. Peer review and critique also help them realize that their opinion is not always the best one, and someone else might be able to improve on it. Teacher Howard comments:

> It allows for a nice rapport where a lot of those kids are used to knowing everything and they have a tough time developing the idea that everything they think isn't exactly right and I think it's allowed them to get comfortable throwing their ideas out there and then accepting "the criticism."

Laptops for communication

With Internet access and an e-mail account (or several e-mail accounts), laptops are a convenient communication tool for students in the one-to-one computing schools. They e-mail their teachers about homework or a question they did not have the chance to ask in the classroom, e-mail their classmates about course projects or personal issues, and chat with their friends using Instant Messenger or other similar programs or in online chat rooms (Lenhart, Madden, & Hitlin, 2005).

Communication with teachers

Laptops provide an additional venue for teacher–student communication. Students visit their teachers' Web sites to find out information about the course or leave a message with the teacher. They e-mail teachers to ask questions they do not want to leave till the next class, to set up a meeting with the teacher, or to ask about assignments or homework if they miss a class. Generally speaking, they can get a response more quickly than with traditional methods. This also provides students more opportunities to ask questions, which is especially true for students who are too shy or uncomfortable to talk to the teacher in person. Without laptops and convenient e-mail access, they would just say, "I'll look it up" or "I'll ask my mom." With e-mails, they can be more "opened up."

In the Crescent Girls High School in Singapore, students sent e-mails and messages directly to their teachers. When interviewed by Radio Singapore International, a teacher commented on the role that Tablet PCs played in her school in facilitating teacher–student communication (Yip, 2004):

> Should they have any concerns, they do not have to arrange any special consultations, to sit down with the teachers face-to-face, to ask them questions, or to clarify their doubts. All the students need to do is send an email, or a simple message, as long as they are within the campus, they can almost get an instant reply from the teacher, wherever he or she is.

This was also reflected in the interviews with both teachers and students in Alpha Middle School. A number of teachers mentioned that they received e-mail messages from students who were too shy to ask questions in the classroom. All students interviewed mentioned that it was easier and more convenient to ask questions or set appointments with teachers through e-mail. One student said:

> [I] talk to them [teachers] a lot more. If you have a question, even if you are at home, you can e-mail them and ask them. You don't have to wait to class to ask. By then you may have already forgot [*sic*]. I can get a response pretty quick [*sic*].

Another student said that it was easier to e-mail teachers to ask questions because:

> We don't have to go and find that specific person. If I am in a different class, I don't have to walk to find that teacher. I can just e-mail the teacher: "Can I come after lunch and redo this test?" or something like that. They can e-mail me back really quickly.

However, due to Internet safety concerns, whether or not students should use e-mail remains a controversial issue and many schools limit or even prohibit students' e-mail use. In schools in the Quaker Valley Digital School District, although e-mail access was given to students, some parents and teachers questioned this decision and complained about an additional burden on them to monitor student behavior (Kerr et al., 2003). Some schools do not allow students to use e-mail. For example, although the majority of the teachers in Maine laptop projects would prefer that their students have access to e-mail, only half of the teachers surveyed reported that their students were allowed to use e-mail, which had an impact on the level of interaction between the teachers and students (Silvernail & Lane, 2004). Similarly, in Michigan's Freedom to Learn (FTL) project, nearly half of the schools did not allow students to use e-mail, which was supposedly one of the simplest ways to support communication, a goal of the project. To maximize the benefit of the program, this policy and others (e.g., allowing student use of equipment in the summer) may need to be re-examined (Urban-Lurain & Zhao, 2004).

Communication with friends

Communication with their peers is a critical element of adolescents' lives. Students at this age love to chat with their friends. They talk during recess, in study halls, after school, and sometimes in class. They talk when they hang out together, talk on the phone, and now chat through their laptop keyboards or microphones. The ways students use their laptops to communicate with and connect to their friends are creative: e-mails, instant messaging, chat rooms, discussion boards, blogs, social-networking Web sites such as MySpace, and online games. Among these, the most popular communication method is the various instant messengers (IM) (with AOL, Yahoo IM, iChat, and MSN currently being the most popular ones). In schools where the use of instant messengers was prohibited, e-mail was the most frequently used.

In Alpha Middle School, instant messengers were introduced by students soon after they received the laptops and immediately started thriving. The popularity of instant messengers lies in the niche they serve: They provide students with a channel to interact with their friends anywhere, anytime. In addition, this high-tech method of interaction was not only fun, but also very fashionable. Students spent so much time chatting through instant messenger that it became a serious problem, especially when teachers discovered that students were spending too much time chatting with their friends in classrooms when they were supposed to be listening to or participating in class activities. It became incompatible with the school philosophy and policies. Consequently, 2 weeks after the laptops were distributed, using instant messengers was banned throughout the whole school.

However, students soon found an alternative method to chat: e-mail. They sent e-mails to their friends and some of them checked their e-mails frequently, even in class when they were "bored." One student reported that instead of going to the next classroom to talk to her friend during recess, "I just e-mail her and say, 'Hi, what's going on?'" E-mail became a messenger and the only drawback was that it was not "instant." It was still very convenient and quick because students checked their e-mail very often and replied to the messages immediately. They did not have to wait until recess to talk to their friends or need to meet their friends in person to talk.

No matter what particular communication method students choose to use, they send each other not merely words, but also other media. In addition to text messages, their e-mails may also have files from school work, a URL of a cool Web site, a song that they want to share with their friends, and pictures they took during a trip, and sometimes they even e-mail their friends a short voice message. When using instant messengers, they use many "emoticons" to express their feelings and emotions. They use flashes to dramatize the conversation. When chatting on an instant messenger, they may just talk to one person at a time, or they may open different IM windows and chat with different people at the same time. Sometimes, they invite friends to join an ongoing conversation, to chat in a group, similar to an online chat room. The conversation can go on and on.

It seems that being able to e-mail and chat makes students feel connected to everyone, so that, as one student put it, "If you have a problem, you can ask your friends to help you online." Students may also use e-mails and online chat to connect with family members. For instance, a student reported talking with her mom online every day since her mom was in another state.

Laptops for Expression and Construction

Designing Web sites

Something about the night sky causes us all, young and old, to ponder over the very basic questions. We are inspired and motivated.

Kalpana Chawla

This quote came from the home page of Johnny, an eighth grader in Alpha Middle School. Johnny was certainly one of those inspired and motivated students. His home page was about the universe: the stars, the sun, the moon, space-exploration-related events, and his favorite sky-gazing technologies. The beautiful night sky not only inspired him, but also motivated him to inspire others. In the "about me" section, he stated:

> For years, I have been interested in astronomy, looking through my telescope at many faraway objects. I have always enjoyed stargazing, but when I was really young, I didn't realize that most people didn't see what we amateur astronomers saw. I wanted other people to see what I viewed through my telescope, so I began to explore astrophotography.

As an amateur astronomer, Johnny had a telescope, an astronomical camera and a few astronomical applications on his laptop. He took his laptop with him to the fields when he observed the night sky, used it with his telescope, and took pictures of the beautiful stars. In order to share what he saw during his exciting night observations, he created a professional Web page to introduce the night sky and share the pictures he had taken. He had to learn computer programming to create a Web page. He was proficient with HTML and Visual C, which he basically learned on his own. And he had a friend to do programming with sometimes. His parents could not help him much on this: "I teach myself. My mom doesn't even know how to turn on a computer, so I didn't get it from her. And she is proud. My daddy just knows about Microsoft Word." He was very proud of himself.

Students like Johnny are not rare. An increasing number of students are creating Web sites, some about their personal interests like Johnny's home page, and some for course projects. Developing a Web site involves much hard work and is knowledge intensive. It requires extensive communication, planning, management, persistence, and sophisticated technical skills. For course projects, students generally work in teams to create a Web site for their group projects. One example of a group project Web site was for the "state project" in Alpha Middle School. In the first year of having laptops, students in the seventh grade social studies class formed groups of four to five working on one of the questions posed by the teacher. These were critical issues facing the state, such as pollution, unemployment, and education budget cuts. Students did research on the topic of their choice, collected data, and suggested solutions for the problems. The final product was a group project Web site for each team. On each Web site, they introduced the topic, explained how they worked on this project, presented statistical data and their analyses of the current situation, and provided their results and suggestions. The Web sites were organized according to the topics and listed in the teachers' course home page accessible to the public.

Writing

Word processing programs make laptops a natural tool for writing. Students type their homework papers, compose essays, and write stories and journals. Writing on computers is much easier for many students than with pencil and paper because they can easily rewrite and edit their work, incorporate images into the text, insert hyperlinks to make their work interactive, and improve the presentation of their final work. All of these can create "a synergistic effect that can potentially deepen and strengthen the intended message of a work" (Bartels, 2000). The research team examining Microsoft's Anytime Anywhere Learning Program indicated that laptops support many useful writing strategies, such as editing and rewriting multiple times, collaborative writing, peer editing, grammar, and spelling checks. All these features not only help students write in a more polished and professional fashion, but also make them take more pride in their work (Rockman et al., 2000). In fact, as pointed out by Warschauer (2006), laptops were used extensively at each stage of the writing process: doing research for prewriting, planning, drafting, revising, and receiving feedback.

A number of laptop projects were set up specifically for improving writing. For example, a special focus of Indiana's ubiquitous computing was to engage students in the high-quality, authentic process of writing, as augmented by technology. The project evaluation suggested that students were more engaged in the writing process when it was enhanced with technology (Lemke & Martin, 2004). The Laptops Initiative for Students With Dyslexia or Other Reading/Writing Difficulties (LISD) in Ireland was designed to help improve the writing ability of students with learning difficulties (Conway, 2005). The Wireless Writing Program in Peace River North in Canada is a one-to-one laptop program devoted to improving students' writing. Approximately 90% of students involved in this program reported that using a word processing program and editing tools was very helpful in improving their work (Jeroski, 2005). Annual program evaluation reports find that, overall, students' writing has experienced considerable improvement in a number of dimensions including organization, meaning, style, and conversations (Jeroski), and the gender gap was closed.[2]

Laptops and other digital word processing and multimedia authoring tools provide students with ways to polish and present their work for authentic audiences. Students with dyslexia and other reading and writing difficulties often are slow to share their handwriting because of its untidy and frequently unintelligible nature. Many participants in LISD in Ireland found that the laptops provided an opportunity to engage with the writing process so that they could edit, revise, and polish their writing for one or more audiences. For example, one student who had produced very little by way of completed texts prior to the initiative because he was ashamed of the poor quality of his handwriting was eager to share his stories with his peers and family. In evaluating the early phase of the LISD, Conway (2005, p. 113) documented the significance of laptops in supporting writing for students with literacy difficulties:

Researcher (R): So what difference does it make to have the laptops?
Student (St)1: If you were in third year you would just be sitting down in the class. You'd have to write stories … you wouldn't be using the laptop.
R: Wouldn't you be writing with the laptop, no?
St1: In third year … if I'd been in third year I wouldn't have been able to use it. Then you'd have nothing to look forward to … when I am writing a story I think that I will write it on the laptop next.
R: Have you shared the stories with anybody else?
St2: Yeah, a few of my friends … teachers … [writing for an audience and developing an identity as a writer]
R: What do they think?
St2: They thought it was good, really good.
R: And did you show them when you had written on the laptop or did you print them out …?
St2: I printed it out.
R: You brought it home?
St2: Yeah, I have a few copies … the teacher has a copy, I have a few at home [printing multiple copies for an audience]
R: Best sellers?!
St2: (laughs)

Creating multimedia products

Researchers find that while multimedia production in regular schools is often limited to PowerPoint presentation, it moves much further and is broader in laptop schools (Warschauer, 2006; Warschauer, Grant, Real, & Rousseau, 2004). Students with their own laptops engage in a wide range of multimedia production activities, including art works, music, movies, animation, and much more. Students who participated in the Cyber Art Project in Hong Kong used the Tablet PCs to facilitate the development of aesthetic sense and their creativity in visual art, and to extend their learning space beyond the classroom (Hong Kong Case Report). During a visit to Beijing, students were working on their artworks about the culture sites:

> But instead of wielding a pencil and a pad of paper to sketch the landmarks before them, the students logged on to wireless-enabled laptop computers and drew on pressure-sensitive digital drawing tablets. Sitting two abreast on the steps of China's Great Wall, which snaked for thousands of miles before them, they sketched the ancient, dun-colored, serrated edifice. (Borja, 2004)

In schools in the United Kingdom, Tablet PCs were viewed as a better tool in painting and drawing artwork than paper and crayon because of the flexible slate and pen, convenience, and immediate effect of students' work. Students reported that they could experiment with their ideas, try out different designs, and make changes with ease (Sheehy et al., 2005; Twining et al., 2006).

In addition to artwork, music, movies, and digital cameras are many students' favorite topics. They like to play with pictures on computers—cutting, copying, rotating, changing colors, and adding special effects. For eighth grader Morgan, Adobe Photoshop is an amazing program because she found that she "could do a lot of things on pictures with it." She saw a friend playing with Adobe Photoshop and started exploring this program on her own and became very skilled with it.

All teens love music. With laptops and the Internet, they can explore their favorite music and singers. They love to listen to music on their computers. Many music programs provide great flexibility for playing music. They can easily control the music they listen to, play back, create CDs, and copy their music from their laptops to their iPod or MP3 player.

With the right software, students can also compose their own music. This happens not only in music class but also in other classrooms. Many music composing programs, such as GarageBand, provide a platform for students to compose their own music and create their own songs. This often requires collaboration with classmates. Students form teams and divide the work according to each member's talents. One student writes the lyrics, another composes the music, and a third, if necessary, may be the singer.

The music they create may be just for fun, or put into actual use in other products, such as a movie. In one-to-one computing schools, many students use iMovie for school projects, especially social studies. For example, when working on solving the problem of an abandoned fast food restaurant across the street from their school, students in Harshman Middle School (Indiana) conducted video-recorded interviews, and then used iMovie to edit interviews and present the final product (Lemke & Martin, 2004). In a U.K. school, some students used Tablet PCs and video cameras in PE lessons to monitor the development of their own work and skills. In another school, teachers and students created video files of lessons to support subsequent revisions and teaching (Sheehy et al., 2006). Some students became so proficient at movie making that they taught their classmates, and even their teachers, to use iMovie (Silvernail & Lane, 2004).

Laptops for entertainment

Computers have become an entertainment center. With games, music, video, and other media capacities, as well as the unique portability and mobility of laptop computers, laptops have become a virtual playground where students can follow their interests; acquire, enjoy, and share entertainment; and explore their potentials. They play on their laptops during breaks, after school, at night, and during weekends. The most popular activities are playing games, chatting (reading and posting) online, and playing with multimedia products.

Playing online games

Playing games has always been a popular computer activity. About 60% of students in our study at Alpha Middle School reported that they played computer games on a regular basis, with a higher percentage for boys (66%) than for girls (50%), and 7% of them reported that the most time spent on computers was on games. The most popular type of game is Massively Multiplayer Online Role Playing Game (MMORPG). Games such as World of Warcraft and Everquest create massive virtual worlds that could compare to continents of our own, allowing the player to choose and customize his or her own persona, all while competing with thousands of other players at the same time. Some students commonly talk of their exploits in MMORPGs, many of which take long hours sitting at a computer. The economies of games such as Everquest and World of Warcraft are greater than those of many countries (Castronova, 2001). They have devoted players willing to sell their virtual belongings for actual cash, playing for hours on end to achieve these goals. So why do they do it? For their self-esteem, just as in real life.

The graphically intensive environments of World of Warcraft and Everquest do not stop at the surroundings. They go into the players as well, creating models for every piece of armor and every weapon. When you start a character and run into a city in World of Warcraft, you are overwhelmed at the number of people with massive glowing swords and powerful, frightening mounts. You look over to that Level 60 Warrior and examine his equipment. As in real life, people admire others in virtual games. This motivates more people to attempt to get those high-level pieces of equipment, playing long hours and adding to the self-perpetuating MMORPG machine. People who are not powerful in real life can express themselves through their virtual characters. There is a sense of great pride when you have reached the highest level attainable in an online game. As in the real world, there is a constant battle to see who is the best; this makes online games extremely addictive.

Playing games is generally viewed as a "bad" use of computers. Adults worry about children being addicted to games and influenced by the violence in some games. Banning games is a common practice in the classrooms and computer labs in K–12 schools. Research on the impact of playing video games on children shows very controversial results: Some studies have reported short-term harmful effects on children, while others have reported considerable therapeutic and educational value (Tapscott, 1998, pp. 164–165).

Do children learn violence from computer games? Tapscott states that playing violent games has always been part of children's play throughout history, from tussling, wrestling, and playing with tin soldiers and fake guns to playing cops and robbers (1998, pp. 162–166). In our study, a teenage boy was asked whether or not he learned violence from the games he played. He said, "No, I actually learned how to avoid violence from playing games." When asked why, he answered, "Because I learned the consequences of violence."

There are other several important benefits gained from playing games. In online gaming terminology, a "buff" is a beneficial spell another character can cast, often improving some attribute of their character for a while. One can consider online gaming as a buff for many. First, online gaming encourages interactions. Normally lackluster students become energetic and excited while playing games with others, going out of their way to assist others, all while trying to improve their experience with other group members. Second, cooperation in teams or groups is an indispensable component for many games. Both World of Warcraft and Everquest allow for guild and group creation, giving gamers a chance to unite in their own little clan and work together to achieve goals. Many places in World of Warcraft absolutely require groups such as these simply because the levels of the monsters are so high. Cooperation is a must, and online games are doing everything they can to support it.

Role-playing in a posting forum

Another virtual playground that has attracted many students is posting forums, which are Web sites specifically designed to let people express their feelings and opinions in writing. Every forum allows a user to create an account, which can post replies to existing threads of messages as well as make its own. In the vast universe of the virtual world, you can find a forum talking about practically everything under the sun, varying from online games to dessert making. Each forum is divided into "boards" that encompass one broad subject, such as a sport. Within these boards are "threads," or smaller discussions that narrow down the broad subject to something more specific, such as how to throw or catch a ball. Within each of these threads are "posts," or messages created in response to the discussion topic. Nowadays, forums allow for picture posting, file sharing, and even a small picture representing the participant in the forum.

To the uninformed, the common interpretation of a discussion forum is a bunch of immature adults and teenagers who log on to make pointless arguments, ending up no further along today than they did yesterday. People think forums are plagued with bad spelling and grammar, full of uneducated imbeciles who have only barely managed to eke out a message and post it. However, RolePlayOnLine (RPoL) proves them wrong. Created by the online persona "jase," RPoL is a forum that allows a user to create an individual "game," or a forum with threads. A normal user can create up to eight games. Along with the various user games, there are also official discussion forums where people can talk about anything on their minds, ask questions about the Web site, advertise for their games, and garner interest for a new one.

A "game" in RPoL is only a game in the loosest sense. Because the Web site only starts the player off with a blank forum devoid of threads, it is up to the creator to decide what the game should be about. As the name implies, RolePlayOnLine is all about roleplaying, with users choosing to become noble knights, daring swashbucklers, desperate soldiers in a war, or even simply high school students. It is completely up to the creator of the game to decide what the storyline will be and what part the players will perform. The creator posts new threads detailing locations, story, and characters—almost like tabletop roleplaying games such as Dungeons and Dragons.

However, each game has a filter that determines how good the writing in one will be. The filter that separates users of RPoL from each other is the creator of the game. To join a game, a potential player has to submit a request to join. The content of these requests is completely at the discretion of the creator of the game—whether he or she wants a name and physical description or a full history and writing sample. This allows various games to cater to many different users' interests and abilities. Therefore, getting accepted into a particularly demanding game with high requirements to join normally yields greater quality writing and more devoted players. Therein lies its value.

RPoL enables the users to learn from being someone else. Each game asks a user to assume a new identity, commonly one of his or her creation or one from a popular movie, cartoon, or book. The players of one game interact in the threads made by the creator of the game, acting out their characters through posts. Thus, the depth of the game also depends on the players themselves, who may write simple sentences or craft entire paragraphs detailing their persona's every thought and action. Players often create images of themselves or model their characters after popular ones in movies, copying their speaking style and writing out their movements. From pretending to be someone else, these people exercise their writing skills, improving them day by day as they encounter other people's writing right next to their own. RPoL attracts a wide variety of people, but it encourages all of them to bring forth their most creative and imaginative writing to contribute to the community as a whole.

Concerns over student laptop uses

It is technically easy to find out what students do on their laptops. However, things become much more complicated when it comes to making the judgment on what uses are "good" and what are "bad." One-to-one computing is still a new phenomenon for many educators and parents. Research on the impact of one-to-one computing is scarce, leaving many issues on laptop use open to question. Consequently, one-to-one laptop projects have been confronted with worries and concerns from school districts, teachers, parents, and sometimes the students. Here we discuss a few of the most frequently raised concerns.

Distractions

One of the major concerns is that having a laptop all the time may distract students. This is reflected in almost all the reports on large one-to-one laptop programs. In a survey study in one of the ubiquitous computing schools, more than one third (39.3%) of the teachers believed that it had become harder for their students to concentrate in class after having the laptops. They thought that students were being distracted by the Internet, e-mail, games, and music. Parents worried that their children's attention span may have been shortened by the various things students could do on their laptops. Students also reported that they liked to surf online, check their e-mails, and talk to their friends through instant messages in class, and some students were really good at "switching programs if the teacher comes."

These problems have received considerable attention from teachers. Teachers at The Maui Campus of Kamehameha Schools in California reported difficulties monitoring student laptop use in classrooms such as students being off task or using laptops when not asked to do so (Rockman, 2003). One teacher at Alpha Middle School said she did not allow students to take notes on their computers, and another teacher required that "the lids of the computers be shut when I am talking." In the Crescent Girls High School in Singapore, teachers could control students' Tablet PCs by locking their screens through virtual classrooms (Yip, 2004). But teachers also expressed the concern on the "level of vigilance required to ensure that students were not using the computers inappropriately" (Jeroski, 2005).

Students' not paying attention in class is not a new problem. As a teacher pointed out, in the past students passed notes in class, while now they chat on their laptops. It is the same issue in a different form. The only difference is that it might be easier for the teacher to catch a note than to detect what is really happening on students' computer screens.

It is also true that students are engaged in some classes but distracted in some other classes. Classroom observations reveal that the extent of students not concentrating in class varied greatly in different classes, as we have discussed in chapter 3. Some teachers had good strategies to monitor students and keep them engaged in their tasks, while some teachers were deeply bothered by this problem. What makes the difference? A student admitted that she talked with her friends in class online "when we are not working on anything that needs my full attention." Another student reported chatting with friends in class "when I'm bored." It is much easier to blame students for not paying enough attention in class than to ask much more complicated and troubling questions: Why do our students get bored easily in class? Why does our classroom teaching not attract and retain students' full attention?

Compared with adults, students seemed to be more optimistic about their ability to deal with laptop-related problems. Most of the students (83.9%) in the study on Alpha Middle School did not think that it was harder for them to listen to the teachers in class after having a laptop. Data from interviews also show that students were able to recognize the potential distractions attendant on having laptops, and they were learning to deal with these problems.

Traditional pencil versus digital pencil

Paul Levinson points out that people have strong attachments to things that they grow up with, including the outmoded media in the last few decades (1998, p. 174). This is also true in the case of adopting digital pencils. Some of the concerns on laptop use are derived from beliefs on what teaching and learning should be. Many people, including teachers, parents, and some students, have a deep attachment to the traditional way of learning with books, paper, and pencil and do not feel secure with technology. For instance, some parents preferred their children to learn from books rather than computers, although no evidence showed that learning from books was better or more effective than learning from computers. This concern is accompanied by a fear of laptops totally replacing books, paper, and pencil. A student said, "I think that losing books altogether is a terrible thing." These concerns influenced the extent to which laptops were being used. For example, some teachers and students were not using their laptops regularly because of parents' and students' preference for hard copies over electronic files (Zucker & McGhee, 2005).

Similarly, some teachers worried about students' ability to read and write with paper and pencil. Like parents, teachers hope students would still value traditional ways of learning through books, paper, and pencil. Their feelings were mixed because, on the one hand, they knew students were going to be living in a digital era where paper and pencil might not be as important as in the past; on the other hand, they still wished students would have good penmanship and appreciate the value of books.

Adults' attachment to the traditional ways of learning might have had some impact on some children. On the U.K. BBC Newsround Web site, a 14-year-old student expressed her concern: "Technology has gone too far already. I think they should stay with traditional methods and maybe have a few more lessons of ICT a week but that is all."[3]

On the same Web site another message posted by a 13-year-old student echoed this concern and complained that "technology is getting more and more complex and it's encouraging young people to forget 'the proper way' of doing things ..."

Gary Stager, a teacher educator and laptop advocate who must have been confronted with the "student-losing-penmanship" question many times, insightfully points out that we should not confuse "writing" (penmanship) with "real writing" (the ability to express one's self) (Stager, 2005). When we focus too much on "the proper way" of doing things, we may forget why we do these things in the first place.

Fear of dependency on computers

Related to the preference for traditional ways of learning is the fear of dependency on computers. For example, taking notes with laptops is one of the most common uses in one-to-one computing environments. As discussed earlier in this chapter, most students loved taking notes on their laptops and greatly benefited from doing so in terms of being more organized, taking better notes, checking spelling and grammar, and sharing notes among classmates. At the least, taking notes on computers greatly improved their keyboarding skill, a competency being taught and exercised in many computer classes. However, some fear that the same spell-checking function that can help students take neat notes may also lead to a decline in student writing skills: "Since the computer does the spell check, the skill of spelling might be low." Even some students think that computers are doing too much for them. For example, a student wrote down her worries in a survey:

> The calculators doing all the math for the kids may be bad because then the kids' brains will not be trained well enough to do equations fast. Also, with the computers doing all the work, the kid's abilities to use his head may not be exercised or developed enough to be able to live without the help of a machine. Basically it's all about getting too dependent on the computers that I am very worried about, and the next generation of kids that might grow up dimmer than the ones before them.

Information literacy

Another concern is related to how to teach students to be more critical and not to take everything online for granted. Teachers worry that students might just copy and paste from the Internet and that they do not think critically and take everything online as facts without careful scrutiny. Many teachers started thinking of ways they could help. For instance, one literacy teacher, through the Media Literacy Grant Program, offered instruction to teach students to be more conscious consumers of the laptops and media, and to learn how to view things critically and scrutinize them. Most teachers we observed gave students specific Web sites that they had evaluated as credible. If students used information from resources other than Web sites provided by their teacher, the resources would be examined by the teacher and some points would be deducted if the resources turned out not to be credible or valid.

Two other concerns that we often hear from the media—students visiting inappropriate Web sites and chatting with strangers—did not appear to be serious problems in the schools we visited or in the one-to-one computing project reports we reviewed.

Summary

Apparently, student laptop uses are more imaginative, creative, and diverse than adults expect them to be. Students use their digital pencils to solve many daily problems they have: doing homework, searching for information on school work, communicating with friends, and developing personal interests. Adults who want to separate learning and play and limit laptop use only to learning activities might be disappointed. For children, playing is a natural part of their life, and they live in both the physical and virtual worlds. The boundaries of these two worlds are blurring. As an increasingly important tool in their real lives, computers are also playing an increasingly important role in their virtual lives.

Meanwhile, "with great power comes great responsibility."[4] The laptop projects have provided excellent learning opportunities and resources, but have also brought some new issues that teachers, students, and parents need to learn how to deal with. Probably the development of one major technology is always accompanied with the loss of some older skills, knowledge, and values. As Bartels (2002) points out:

> With the development of books, students lost some degree of fluency with their memorization and verbal skills. With the development of electronic information, students will lose some degree of fluency working with information on paper. In both cases what makes the loss acceptable is the enormous gain in access to vast and rich new realms of information.

Laptops versus handhelds 5[1]

One day, 2 or 3 billion people will have cell phones, and they are not all going to have PCs ... The mobile phone will become their digital life.

Jeff Hawkins, inventor of the Palm Pilot

Hundreds of millions of people are not going to replace the full screen, mouse and keyboard experience with staring at a little screen.

**Sean Maloney, an executive vice president
at chipmaker Intel (Stone, 2004)**

Introduction

Although laptop computers have been a dominant device for most large one-to-one computing projects, especially those in the United States, Canada, and Australia, they are by no means the only viable one. Tablet PCs, for example, are a popular choice in the classrooms of the United Kingdom, Singapore and Hong Kong (e.g., Sheehy et al., 2005). Many small digital devices have also entered classrooms years ago. Calculators are widely used in math classrooms in the United States. Electronic English dictionaries are a common digital tool for students and some adults throughout Asian countries (Chan et al., 2005). Recent years have witnessed a rapid growth of emerging technologies that can be used in classrooms. Now there is the $30 LeapPad[2] interactive learning system, the $100 Pentop computer,[3] the $140 graphing calculator, the $200 iPod, the $300 Palm Pilot (or a second-hand one at $50), or other similar handheld devices that are commonly known as personal digital assistants (PDAs).

These devices vary not only in their size and cost, but also in their functions. Some can perform a set of very focused but limited functions (e.g., LeapPad and graphing calculators); others have more generic functions, such as the PDA. These devices are not the same as general-purpose computers, but nonetheless provide certain mobile and portable computing capacities that can be harnessed to support teaching and learning, at an affordable price. Price is not the only advantage these alternative devices have over laptop computers. Handheld devices are even more portable, easier to carry, and simpler to operate than full-blown laptops. Thus, they have become an increasingly attractive alternative to laptops for one-to-one computing initiatives (Choi et al., 2007).

In this chapter we review the current situation of other digital devices in schools, examine their advantages and disadvantages, discuss conditions that influence the adoption of handheld computing devices in schools, and tell an in-depth story about the implementation of PDA projects, comparing them with laptops.

Possible alternatives to laptop computers

With the large variety of information and communication technology (ICT) devices currently available and new ones emerging continuously, it is impossible to exhaust the list of digital devices that can be used for educational purposes. Here we discuss a number of the most commonly used ones, categorized by the range of functions they can perform.

Computing devices with generic functions

Other than laptops, PDAs are the most popular choice of one-to-one computing projects. Given their small size, portability, and comparatively affordable price, PDAs are reaching increasing numbers of schools. As early as 2000, Consolidated High School District 230 in Illinois bought 3,000 Palm handhelds to use in three high schools.[4] According to the 2004 Quality Education Data, 55% of American public school districts used PDAs in the 2002–2003 school year, with an additional 8% expected to purchase them for use during the 2003–2004 school year (Honey & Culp, 2005). Many schools participating in the Michigan Freedom to Learn (FTL) project equipped teachers and students with PDAs (Urban-Lurain & Zhao, 2004). The Palm™

Education Pioneer (PEP) program provided PDAs to 102 classroom teachers and their students (Vahey & Crawford, 2002). In the United Kingdom, after a pilot project that provided PDAs to 150 teachers and 100 students, some schools and local education authorities (LEAs) are seriously considering equipping all students with them (Perry, 2003).

In terms of the availability of handheld computers by grade level, secondary schools are more likely than elementary schools (14% compared with 9%) to provide handheld computers to students or teachers for instructional purposes (Parsad & Jones, 2006). This may be because of concerns that younger children lack the necessary technology proficiency and motor skills to use handheld computers (Vahey & Crawford, 2002, p. 8). However, the PEP program evaluation report conducted by Stanford Research Institute (SRI) finds that young children in elementary schools are able to use handheld computers, and there is no difference between secondary schools and elementary schools in terms of the frequency of handheld use, the type of use, or teaching style. Furthermore, elementary school teachers are found to be more enthusiastic about using handheld computers than secondary school teachers.

The authors suggest that this might be because the simpler content of elementary school classes is more suitable for the small handheld screen than that of higher grade levels, and elementary school teachers are less constrained by standardized tests and thus had more flexibility in using these handhelds (Vahey & Crawford, p. 9). Different from the American practice, the pilot PDA project in the United Kingdom includes more primary schools than secondary schools (18 vs. 9), and no grade level difference is reported in the evaluation report (Perry, 2003).

Communication devices

Cell phones probably are the most popular ICT devices worldwide. In the United Kingdom, 70% of the general public and 90% of young adults have cell phones (Naismith, Lonsdale, Vavoula, & Sharples, 2005). In China, more than 459 million people owned a cell phone in 2006, and this number had been growing at 50 million annually (*Xinhua English*, 2006). In the United States, more than 203 million people have cell phones (Leo, 2006); among them, more than 60 million are teenagers (Youra Studio, 2006). This number will certainly grow because cell phones have broken into the top 10 most wanted children's gifts worldwide (BBC News, 2006a).

Young people use cell phones to make phone calls, take pictures and videos, play games, and, most importantly, send and receive text messages. In the United States, about 63% of Americans from ages 18–27 text message (Rainie & Keeter, 2006), and about 32.5 billion text messages were sent through cell phones in the first 6 months of 2005 (Youra Studio, 2006). The popular use of cell phone text messaging has created a "thumb generation"—young people who use their thumbs extensively for text messaging and playing games (Digital Europe, 2003).

The wide adoption of cell phones' multiple functions provides great potential for various educational uses. For example, the U.K. BBC Mobile Bitesize project offers small packages ("bite size") of quizzes and practices on several subjects to students' cell phones to help them work on their General Certificate of Secondary Education (GCSE).[5] Over 650,000 GCSE students have used this program (Vavoula, 2005). In September 2005, China Farmer University initiated an "Instant Messaging Learning Platform."[6] This platform sends instant messages, free of charge, to the cell phones of thousands of farmers dispersed in fields and farms around the country. Farmers can also send messages back to the platform, asking questions or making their suggestions. These interactive and instant short messages help farmers learn practical skills and knowledge about farming, processing, managing, and marketing, and keep them updated with the latest farming trends and most current market information.

Handheld devices with more specific functions

Some handheld computing devices have more specific functions and are used extensively for a limited number of purposes or a specific subject, such as calculators for math, electronic dictionaries for language learning, and special devices for writing and reading. In 2000, more than 80% of high school mathematics teachers in the United States used handheld graphing devices in their classrooms (Keefe et al., 2003). AlphaSmarts, a type of portable writing device, is capable of running basic word processing programs that allow students to compose, edit, cut, copy, and paste text; save; print; and perform spell-checking. It is used in approximately 40% of American schools, and the total number of the devices in schools has reached 1 million (Russell, Bebell, Cowan, & Corbelli, 2002). Leapfrog's Leappad offers interchangeable books and cassettes that allow the user to read, listen to stories, and participate in interactive games and activities to support active learning (Romig, Yan, & Zhao, 2004). It is reported that more than 40,000 classrooms in the United States use the LeapFrog programs (Boone & Forsythe, 2004). These specific devices have been found to help students to improve their writing and language skills (Vahey & Crawford, 2002, p. 11).

Gaming devices

Playing games has become part of modern living. More than half (56%) of America's 8- to 18-year-olds now live in a home with two or more video game consoles, and nearly half (49%) of this population have one in their bedroom (Rideout, Foehr, & Roberts, 2005). Educators have started exploring the potential of using readily available game consoles for learning. For example, Skills Arena (Lee et al., 2004), a math video game implemented using the Nintendo Game Boy Advance system, is used to help students learn traditional math curricula (Naismith et al., 2005).

As we have discussed elsewhere in this book, the convergence of functions is one of the major trends in ICT development. Soon it will be increasingly difficult to categorize ICT devices by their functions. Computers can be used to make phone calls; cell phones can be used to surf the Net or e-mail; and PDAs can perform calculations, compose text, and make phone calls as well. A new mobile phone device recently released by Apple Computers, iPhone, is similar to a touchscreen computer, with high interactivity and connectivity.

How are these devices used?

Existing reports on how handheld computers can be used in schools have been largely positive and optimistic:

> For the first time, students can have a truly portable and personal low-cost anytime/anyplace general learning device that can be used in any number of individual or collaborative learning activities such as taking measurements at a stream, learning vocabulary words while waiting to be picked up after soccer practice, or working on a report while on a long car ride. (Vahey & Crawford, 2002, p. 5)

Although we do not know if students would want to learn vocabulary words in the soccer field after a practice game, it is true that these portable computing devices, similar to laptops, can be used in many ways for different learning activities. In a report that reviews the mobile learning literature, researchers at the Futurelab conclude that mobile technologies are being used to support six types of learning and teaching activities: behaviorist, constructivist, situated, collaborative, informal and lifelong, and learning and teaching support (Naismith et al., 2005).

In terms of content, handheld computing devices are used in all subject areas, especially in science, math, language arts, and social science classes. The mobility of handheld computers and the availability of calculators and spreadsheets make them easy to use for data collection and analysis in science and math classes. In the final evaluation report on the PEP program conducted by SRI, well over 90% of teachers who engaged in science-based curricular areas reported that handheld computers can improve the quality of learning activities (Vahey & Crawford, 2002, p. 9).

In math classes, students use PDAs in ways comparable to graphing calculators, using applications such as Match-My-Graph to work collaboratively to facilitate and communicate their understanding of math graphs (Tatar, Roschelle, Vahey, & Penuel, 2003). In an advanced physics class, students used PDAs on a daily basis. They collected data from the experiments they conducted in their lab, checked syllabi on the Web (using AvantGo), beamed experimental data back and forth, collected notes and information, collaborated with classmates, and wrote lab reports, as well as kept up with daily news (Tartar et al., 2003, p. 43). In another classroom, students used Sketchy to design an animation that required them to decide how to represent physical phenomena, causal processes, space, and time—all integral to understanding science—and then shared their work with the teacher and their classmates (Tatar et al.). In addition to being used for teaching and learning purposes, PDAs are also being advocated as an assessment tool to facilitate broadening the range and frequency of teachers' assessment of inquiry in science (Penuel & Yarnall, 2005).

For language classes, handheld computing devices are handy and useful tools. In English classes and English as a second language (ESL) programs, students use dictionaries, writing and reading programs, and exercise programs on their handheld computers. They can read stories from their eBooks, look up words from a dictionary or a thesaurus, and use a Memo Pad or word processor to write an essay. For foreign languages such as French, students use dictionaries, grammar exercises, vocabulary programs, and quiz applications (Dell, 2003). A majority of English teachers participating in the PEP project (85%) used PDAs in student learning activities and reported that handhelds improve the quality of the learning experiences (Vahey et al., 2002).

The easy text input and output functions of cell phones also make them a convenient tool for language learning. A southern Italy organization, for instance, designed a basic Italian language course that can be delivered to learners' mobile phones through text messaging. Students learn the language concepts and practice dialogues. They can also take quizzes on their phones, send their responses to the course system, and receive feedback and suggestions for improvement (Attewell, 2005).

Even in schools with one-to-one laptops, other portable devices such as PDAs and graphing calculators can be a valuable addition. For example, handheld calculators were extensively used in Mr. G's math classroom in Alpha Middle School. His students used these calculators in many ways, such as programming, drawing, calculating, and graphing, and they also learned how to interpret these graphs. He liked to use the handheld calculators for a number of reasons. First, in his class, every student had one handheld calculator, so it was easily accessible and he could involve the whole class in the learning activities. Second, he had almost never had any technical problems with these handheld calculators, so he and his students trusted these devices as being reliable and able to "always work." Third, he was very proficient in using the calculator, both with the hardware and the program, so it was easy to use, to teach, and to answer student questions. In addition, he found that students were fascinated more often with the calculators than with their laptops because, according to Mr. G:

> The computer is expected to do all these great things, but the calculator, they just expect it to add, subtract, multiply and divide. So if we do something interesting on the calculator, they are like: "Look at what the calculator can do!" although they know their computer can do the same thing, even better and more colorful[ly].

This unexpected comparison adds to students' motivation in using the calculators (Lei, 2005).

Mobile devices can be used in participatory simulation games in which players actively play a part of the game in a physical setting

(Vavoula, 2005). For example, Klopfer and Squire (2004) used PDA devices to engage students to play a virus game, using PDAs to track players' interactions, to learn about how a computer virus spread. In the savannah game designed to learn lion behavior, students played lions in the savannah in a specified field, each carrying a PDA that was tracked by GPS to tell them information about their location and situation (Facer et al., 2004).

Research studies on the impact of these handheld computing devices report several benefits to students. For instance, in a study examining the effectiveness of learning with the LeapPad system, it is found that first graders who used this learning system show significant advances in phonics, oral reading fluency, and retelling than students in the control group (Romig et al., 2004). According to the SRI PEP evaluation report, teachers who use the personal use strategy view handheld technology as conferring relatively greater benefits to students in a number of areas: increased time spent on schoolwork outside school, increased organization in general, increased initiative in finding ways to use the handheld computer for school or learning-related tasks or activities, increased time spent on voluntary (not assigned for school) learning activities, increased homework completion rate, and increased opportunity to use technology (Vahey & Crawford, 2002, p. 38). Russell and his colleagues (2002) found that having their own AlphaSmarts changed the way students approached writing, improved the quality of their work, and, at the same time, encouraged a greater sense of student ownership, responsibility, and empowerment.

Handheld computing devices are found to be natural learning tools to facilitate collaboration and communication (Naismith et al., 2005). Even for very young children, handhelds can enhance social interaction and learning. In a study of handhelds in teaching mathematics to 6- and 7-year-olds in Chile, Zurita, Nussbaum, and Salinas (2005) identified at least three benefits for learning: enhanced social interaction, greater motivation, and improvement in basic skills. In addition, for students without much computer experience, PDAs may help them build their confidence and familiarity with new technologies (Garden Valley Collegiate, 2002). Teachers also benefit from handheld computing devices. For example, they can use the data administration capabilities to record attendance, keep meeting arrangements, and organize their lesson plans (MOBIlearn, 2005).

The good and bad of other devices

Advantages of handhelds

Compared with laptops, handheld devices have a number of advantages that contribute to their growing popularity. The first advantage is their relative low cost. For schools that hope to make sure students have access to a computing device all the time, handheld computing devices seem to be an attractive alternative to laptops because it is generally less costly to equip students with their own handheld devices than with a laptop (MOBIlearn, 2005). For example, the price of an AlphaSmart is around $200, which is about one fifth of that of a regular laptop. PDAs cost approximately the same. The low cost of handheld computers is certainly a factor when schools make technology decisions. In Joyner's report (2003), a technology resource teacher at Ohio's Green Middle School, which rotated mobile handheld labs among classrooms, commented on the difference between laptops and PDAs in terms of cost: "You can buy wireless laptops for $2,000 or Palm handheld computers for $200. Where's the dilemma?"

Another reason for the increasing popularity of handheld computing devices is the wide range of tasks they can perform. At about one fifth of the cost of a laptop, PDAs have the essential functions for many learning activities. The manufacturers even claim that teachers and students can accomplish about 80% of the tasks they can do with a regular computer (Joyner, 2003). In fact, portable computing devices vary greatly and many of them have specific functions. Although they do not have the full range of flexibility of the computer, these devices have many of the necessary features for the intended learning tasks. For example, handheld calculators can be used for collecting and analyzing data and presenting results. For writing, the AlphaSmart can work better than laptops. An online review written by a parent commented, "At only 2 pounds, and for the price in comparison to a laptop, if you're just looking to do some writing, this AlphaSmart is perfect!"[7]

A third reason that makes handhelds appealing is that they open up new possibilities for learning, especially collaborative learning (Zurita & Nussbaum, 2004). Handhelds with wireless capacity provide opportunities for learners to interact with each other both face to face and wirelessly, leading to enhanced social interaction and greater motivation to learn (Nussbaum & Zurita, 2005; Zurita et al., 2005). Wireless handheld devices have the potential to provide the kinds of collaboration and social interaction that have been associated with enhanced learning, and as such they fit well with the social constructivist approaches to learning (Cole, 1996; Vygotsky, 1978). Consistent with this social constructivist perspective, Klopfer, Squire, and Jenkins (2002) argue that PDAs afford greater portability, social interaction, context sensitivity, connectivity, and individuality than alternative digital tools.

In addition, with smaller size and longer battery life than laptops (MOBIlearn, 2005), handheld computers are more portable and durable, allowing teachers and students easily to bring computing technology with them into field study areas and on field trips. Vavoula (2005) argues that mobile devices enable students to engage in situated learning because they can take their devices outside classrooms to the fields or contexts related to their learning content.

Disadvantages of handhelds

While handheld computers provide students with many new opportunities in classrooms, they also pose some challenges to schools. For instance, unlike a desktop computer, whose functionality is by now familiar, handheld computers require their users to learn new ways of operating them (special handwriting strokes), and the charging and downloading procedures are different. Based on existing studies and reports, the major challenges are discussed in this section.

Lack of support

The first challenge is related to the support for the use of handheld computers, including supporting infrastructure and technical support. Although all handheld computing devices can be used independently, to take full advantage of their functions, many peripheral technology devices are needed. In the PEP program, the successful use of PDAs was supported by many peripheral technologies such as keyboards, digital cameras, and probes (Vahey et al., 2002). Similarly, many other portable computing devices also need supporting infrastructure. For example, most handheld computing devices do not have print-out capabilities unless connected to the network; therefore, a network and printers or IR-enabled printers are needed for teachers and students to print out their work.

All technologies can and do break, and thus need timely technical support. However, technical support for handhelds is scarce in most cases because, although handheld technology has been widely used as a business tool for several years, as a learning tool it is still new in schools. Many districts and schools are unable to provide sufficient technical support on PDA use (Vahey & Crawford, 2002, p. 45). Most teachers do not have much experience with PDAs, so it is not easy to obtain help from colleagues. Teachers in the United Kingdom's PDA pilot program reported a lack of technical ability and familiarity with the PDAs as well as a lack of a clear vision of what the PDAs' potentials might be (Perry, 2003).

If handheld devices break down often, it is difficult to integrate them into daily practice. For example, about one fifth (21%) of teachers surveyed in the SRI PEP evaluation study reported that the procedures and time taken for charging handhelds, replacing and supplying batteries, and the loss of student work due to charger loss were a major problem in using PDAs (Vahey & Crawford, 2002, p. 58). Teachers in Michigan's FTL program reported that it was difficult to integrate the PDAs because the keyboards were not working properly (Urban-Lurain & Zhao, 2004).

Lack of appropriate learning content/application

Many computing devices, such as PDAs and cell phones, were not originally designed for learning purposes and do not have many learning applications, so learners can only use whatever tools are available (Waycott, Jones, & Scanlon, 2005). Consequently, it is often the case that teachers and students receive these technology devices but do not know what to do with them. Teachers often need to spend time to research available software and applications, evaluate their appropriateness for teaching and learning, and ensure resources for their use (MOBIlearn, 2005). In the SRI PEP project, teachers noted that there were too few education-specific applications (Tatar et al., 2003). They reported that having the appropriate software and peripherals was key to the success of their handheld uses. All teachers believed that "providing teachers with instructional materials and resources that integrate handhelds into the classroom" was an important factor for successful integration of handheld computers, with 75% of them strongly considering it a "very" important factor (Vahey & Crawford, 2002).

Running cost can be high

The initial purchase price for most handheld devices is much lower than that for laptops, but the running cost can be quite high. For instance, while it costs less than $40 for a LeapPad learning system, additional books, at about $10 each, are required for real learning to happen. Cell phones enjoy low initial cost. Most phone companies offer free cell phones with service contract. But the service fee can reach $700 per year (Houser, Thornton, & Kluge, 2002). Thus, if not everyone in the class already has a cell phone, learning activities cannot be organized around cell phones. Wireless connection fees for cell phones and PDAs are also expensive, which limits the use of some key features of these handheld devices.

Limited capabilities

The small size of most handheld computing devices provides good portability, but, on the other hand, also limits many important functions. For example, the small screen size affords very limited display capabilities, making it ineffective to display some files or Web pages (Vavoula, 2005). Most handheld devices have limited input options. The data input rate is much lower than that on laptops (Houser et al., 2002), making it time consuming to input data, especially free text (Perry, 2003). In addition, even if handheld devices can be connected online, limited bandwidth offers limited learning content. Thus, content for handheld devices has been mostly reading and multiple-choice quizzes (Houser et al., 2002). As a result, "bandwidth is not yet good enough for substantial online learning" (Attewell, 2005).

Concerns over inappropriate use

Like laptops, handheld devices can cause concerns too. One frequently cited example is the inappropriate use of the infrared beaming function. This function enables students to send files to other similar devices and is a highly valued feature because it can facilitate communication and collaboration among the students and with the teacher. However, it can also be used for other purposes. Adults worry about the possibility that the beaming capability may affect order and integrity in the classroom. For example, students may beam inappropriate content to each other, download illicit content, or cheat in exams (Tatar et al., 2003). In New York City, because of serious concerns over student misuse of cell phones, such as cheating, taking inappropriate photos in bathrooms, and organizing gang rendezvous, cell phones were banned in the whole 1.1 million-student school system (CBS News, 2006).

Handheld devices can also create distractions in classes. Many students used PDAs to play games. In the SRI PEP evaluation study, controlling or restricting students' use of handhelds to on-task activities or functions was the management task that teachers reported having the most problems with (Vahey & Crawford, 2002, p. 47). Similarly, in the project evaluation report (Urban-Lurain & Zhao, 2004) on the Michigan FTL program, distraction caused by playing with PDAs in class was reported as a serious concern of teachers. One teacher whose school received PDAs discussed his frustration:

> The potential for handhelds to enhance education is fabulous; the reality is anything but [...] I found them to be a tremendous distraction from the educational process and thus a huge frustration for me. It was like having a room full of Gameboys. It is very difficult to enforce proper use, since you have to be almost on top of the screen to see what the student is doing, and when the teacher approaches, students will exit out of their games. It soon also became very difficult even to get students to bring their handhelds to class, since they weren't interested in using them when they found out that they weren't allowed just to play games on them.

However, there is disagreement in existing studies on whether or not having handheld computers causes distraction in classrooms. In the study conducted by Tatar and her colleagues, none of the teachers complained about disruptive behavior in classes where students used handheld tools. On the contrary, teachers reported decreased disruptions compared to classes without these tools. Researchers also noted a decrease in natural behaviors such as going to the bathroom or getting a drink of water (Tatar et al., 2003). The difference in the extent of disruptive behaviors caused by having handheld computers in different classrooms is certainly an interesting phenomenon and worthy of further investigation.

Researchers suggest that educators should not just view handheld devices as disruptive to classroom activities, but rather need to explore the potential of these devices and find ways to put them to good use to facilitate teaching and learning (e.g., Naismith et al., 2005; Sharples, 2003).

How the positives and negatives played out in one study

The advantages and disadvantage of different ICT devices need to be taken into consideration when making technology decisions. Schools must carefully evaluate their needs and existing technology resources, choose ICT devices that can best fit in their school system, and provide sufficient support to facilitate the meaningful integration so as to achieve their educational goals.

Recently, we were able to observe the introduction of handheld computers in schools, and thus we had the opportunity to examine how the advantages and disadvantages of handheld devices play out in different local contexts. In 2002, the state of Michigan, through the FTL program, granted money to 15 school districts to support their acquisition of wireless learning technologies. Districts were allowed to elect what type of equipment to use, and they made a variety of selections of different types of technology from different manufacturers. Some districts elected to use only laptops and others only handheld computers; some chose to use both laptops and handheld computers.

In 2003 when researchers surveyed school personnel involved in the program, it was found that, in general, laptops were much more successful and popular than PDAs. However, it was found that another school district had successfully implemented a PDA project, so the failure of PDAs in the FTL program schools did not necessarily mean that PDAs could not survive in schools. As further research was conducted, it was clear that a variety of factors involving the school settings and the devices themselves affected the implementation of handheld computers. In this section we discuss these factors by first comparing the implementation of PDAs with that of laptops within the FTL program, and then PDAs in the FTL program with a non-FTL school district that successfully implemented a PDA project.

PDAs versus laptops: the survival of two different technologies

As researchers learned more about the use of these two technologies in the schools participating in the FTL program, it was found that handheld computer users reported less satisfaction than their laptop-using counterparts. Moreover, most of the schools using handheld computers encountered serious problems, and many of the teachers had abandoned using the handhelds. Thus, the question arose as to why the handheld computers struggled to survive in these schools. To address this question, researchers interviewed 10 handheld computer users and 8 laptop users from a total of 16 different schools. Conversations in these interviews revealed a complicated picture of how and why PDAs were underused or unused in these schools.

The ecological framework discussed in previous chapters can provide a number of explanations of the finding that laptops survived much better than PDAs. First, the laptops are a familiar species. Teachers and students already know how to use a computer, so it is easier for them to integrate laptops in their teaching and learning. In contrast, PDAs, as a new species, break the system equilibrium and create disturbances, which may cause emotional uneasiness. Second, the school system did not have sufficient and appropriate resources to support the growth of this new species. Although many teachers mentioned that they experienced some types of problems regardless of whether they had used handheld computers or laptops, what was striking was the depth and pervasiveness of the problems that people experienced using handheld computers. This is likely because schools have resources to support the use of laptops, such as printers, projectors, and computer software, but most schools lack resources for and expertise in PDAs. In addition, PDAs cost more resources because teachers and students need to learn how to use them, which requires more time—something that is always in short supply in schools.

Third, competition between different technologies can be intense, and PDAs do not have a strong edge in the competition. Teachers make their technology decisions based on cost-benefit analysis (Lei, 2005; Zhao & Frank, 2003). PDAs hold lower perceived value than desktop and laptop computers, but cost much more time, effort, and other resources than traditional technologies such as blackboard or paper and pencil. It is therefore difficult for them to conduct cost-benefit analysis favoring PDAs. Overall, compared with laptops, PDAs consumed more resources and were perceived as less beneficial than laptops, and consequently they did not survive as well as laptops.

PDAs versus pdas: a tale of two introductions

However, the preceding comparison of laptops and PDAs does not mean that the latter cannot survive in schools. In fact, another school system (BTS) successfully introduced handheld computers over the past 2–3 years. The experiences of this district were almost the polar opposite of those of the FTL participants. Analyzing the experiences of these two groups of handheld computer users and their very different outcomes further illustrates the complex and dynamic process of technological adoption and provides some important guidance to future attempts to infuse new technologies into schools. A comparison between the implementation of PDA projects in the FTL schools with implementation in BTS schools revealed a number of reasons for PDAs' lack of success in the FTL project schools.

Technical problems: functional versus nonfunctional

The first major reason is technical problems. When talking with FTL handheld computer users about problems they encountered, researchers were surprised by the fact that they had so many technical problems. Over and over, people spoke of keyboards that malfunctioned out of the box, batteries that would not hold a charge, memory that was insufficient, and continual hard-reboots, where children would lose all their work, as expressed in the comments of two teachers:

> The keyboards were a disaster. The first ones we had broke before we even got them out of the box. We got new ones and they never really worked well. I never abused mine and I took it to a meeting to show to our school board and I'd only used it once or twice and it didn't work. The same happened with the kids; they would push one key and another letter would come up. That's why I gave a bad grade to the equipment and to the typing because for the most part they couldn't type reports.
>
> As time went by, more and more [Palms] became unreliable. It might have been more than 12 [that he sent back]. The kids didn't trust them. One boy was crying; he had spent 3 hours writing a paper the night before and couldn't get his paper up. I said to the heck with them, it's not worth it.

Handheld computers are widely used among individuals without such complications. Why did FTL have so many problems with PDAs when BTS did not?

It was as if a genetically defective strain of handheld computers had been introduced in these districts; few if any of them were fit enough to survive. Upon investigation, it was found that to cut costs some of the districts had opted to purchase refurbished handheld computers and keyboards. They had essentially selected a strain of species that was superior in terms of price but inferior in terms of function. The sad fact of the matter was that this had disastrous consequences. One of the schools sold off its handheld computers. Teachers in other schools reported that they stopped using them in the second year of the program, deeming them too much trouble and too unreliable to use. High levels of unreliability led teachers to believe that any exploration with the devices was a waste of time, and thus they rejected the use of this tool.

Most of these technical problems were avoided in the BTS school district because the PDAs they purchased were brand new and functioned well. Teachers and students did not have to deal with the frustration of broken keyboards or malfunctioning PDAs. After exploring with the PDAs for a period of time, teachers were able to use them in their classrooms to support teaching and learning activities. In this case, the functional PDAs were able to find a niche and meet teachers' and students' needs, thus surviving the selection process. The BTS technology coordinator spoke of such selective processes as well. She mentioned another district (HVS) that purchased two different brands of handheld computers. One type was an abysmal failure, having inordinate numbers of technical problems. The other type functioned quite well and was much sought after. At the end of this project, HVS abandoned the malfunctioning handheld computers, unable even to sell them off.

The positive experiences of BTS and the negative experiences of FTL and HVS illustrate that the functionality of a new technology is critical to its survival in a new environment. If the new technology causes the waste of time and resources and does not serve teachers' needs, it will not be incorporated into the school ecosystem.

Time to adapt

It takes time for a new species to take root and thrive in an ecosystem. Likewise, it also takes time for a new technology to be successfully integrated in schools, because both teachers and technology need to go through an adaptive process, and teachers need time to negotiate the complicated task of adaptation and integration of technology. If teachers are given the time to adapt to the new types of technology in their classrooms, they have a better chance to build positive experiences with a technology, which makes them more likely to adopt its use (Zhao & Frank, 2003). One of the major differences between the handheld computer system in FTL schools and that in BTS is the timetable of the two efforts to establish handheld computers in schools. In the BTS model, a very gradual approach was taken to introducing handheld computers. As their former technology coordinator said:

> It's a complex picture to do on a large scale. Part of our success is that we started it out on a small scale—we started with one class, and then we went to a grade level, then we introduced it to another grade. We didn't give them to the kids right away; the teachers used them for a year first, and then we gave them to the kids. All building staff were issued handhelds and had professional development on ways to use their handhelds ... If the teachers don't have the comfort and confidence, they will struggle with it. If they know how to do the basics, and feel comfortable, then they'll be able to show kids what to do and be more successful.

By contrast, the picture that people in the FTL program painted was very different. Handheld computers were delivered to schools during the spring semester of 2003, and teachers had a matter of 3–4 months to learn to use handheld computers and integrate them into their teaching. Across the board, teachers commented (in interviews) on the short time frame of the FTL roll-out:

> The timeline that was dictated to [our intermediate school district] by FTL was so unrealistic. [The ISD] didn't have time to adequately prepare to execute this kind of a program. The professional development, because we had no time, was all dumped on us midyear. There was no possibility to schedule professional development. In other words, our school calendar gets set for the year, and at the beginning of the year it didn't include FTL.
>
> The only major problem that we had was the time span, the amount of time we had from the time we got the equipment and the time we had to start the program and end the program. It seems like it went too quickly. It would have been nice to have at least 6 months for the professional development for teachers. We had teachers who were afraid to touch the computer, let alone the

handheld computer. We were a little bit behind when we started. The teachers worked hard, but we were behind because of time ... Too much, too fast ... Next time, please slow it down.

With the rapid onset of the FTL program, many participants struggled to make use of their handheld computers. Teachers and students wrestled with how to download from their handheld computers to a common source so that teachers could review student work. Providing a method for students to charge their handheld computers was also a challenge because many districts hesitated to send the chargers home with students due to the fact that the cradles had cables necessary for downloading information. All of these problems were worsened by teachers' lack of experience with handheld computers; they just did not have the time to adapt to using handheld computers for teaching:

> The only [other] problem was that I am a novice and when there are any problems with technology I don't know how to fix them. Then it becomes a waste of time in the classroom. If a kid didn't know how to do something, I would have to send them to someone to figure it out. That was a source of frustration for me, not to know how the hardware works and the thing that I was taught to do doesn't work; I didn't know what to do next. I didn't know how to intuit what to do next. (quotes from teacher interview)

By comparison we see clear difference in the two school districts. In the BTS model, teachers and the rest of the school system were given time to adapt. The slow pace of introduction allowed for the development of procedures to work out the kinks. Teachers became familiar with the handheld computers and their uses. Administrators tested that file-sharing systems worked smoothly. The BTS approach ensured that when the time came to use the technology with children, hardware systems and use protocols were up and running. In contrast, FTL handheld computer users were given a drastically shorter timeline. Teachers were not able to build up the positive experiences and knowledge that are fundamental to integrating a new technology. As a result, they had inadequate time to go through an adaptation process. What little adaptation time there was occurred "live" while teachers were in front of their children during class time, which is hardly an ideal time for problem solving, as teachers suggested earlier (Urban-Lurain & Zhao, 2004).

The scope

The third issue, related to the second one, is the scope of the FTL implementation. Handheld computers were delivered to FTL schools in the spring of 2003. One middle school alone received 220 handheld computers for all its seventh grade students at once. Without sufficient preparation and support, this led to an environmental overload in their schools. When problems occurred, they were often overwhelming. One teacher commented:

> We have an instructional aide that dealt with all the problems. There was an insurmountable problem with 220 Palms; at the start of the day there would be so many kids at her desk trying to get their programs fixed so that they could use them during the day.

The difficulties FTL handheld computer users experienced were ones that would perhaps have occurred in any event, but given the fact that the system was overloaded by the new technology, the equilibrium of the school environment was severely disrupted. Teachers began to question the use of the handheld computers. One school technology director gave up the struggle: "I said to the heck with them, it's not worth it," and, with the administrator's consent, he then sold off the school's supply of handheld computers to interested students.

Overload, combined with a short time frame, made it difficult for FTL teachers to adapt to and incorporate handheld computers. Because some teachers rejected handheld computers, they were essentially selected out of the school system.

By contrast, BTS started from a small scope and gradually expanded it. This type of trickle introduction reduced the stress on the entire ecosystem, allowing the infrastructure, including hardware systems, use protocols, and teachers, to adapt to and incorporate the new technology.

Summary

Handheld computing devices, like other digital tools discussed in this book, are becoming increasingly embedded, ubiquitous, and wirelessly networked in schools. They can be a reliable and affordable method of realizing the vision of ubiquitous computing-enhanced learning, especially for schools with limited budgets. Nevertheless, the successful adoption of handheld computers depends on a number of factors that must be taken into consideration when making technology decisions (Naismith et al., 2004; Zurita & Nussbaum, 2004). First, the many types of handheld computers differ in what they can do and how much they cost. Schools should choose the ones that can best fit their needs. Cost is a major concern but cannot be the only concern. The most expensive ones are not necessarily the best option and neither are the cheapest ones. Some are cheap to purchase but very expensive to maintain, such as the refurbished PDAs in the FTL program. Therefore, there has to be a balance between short- and long-term costs.

Second, handheld computers, no matter how "small" they are, need effective and timely technical support in order to be integrated into classrooms. Third, as in the gradual co-adaptation between a new species and the ecosystem, it takes time for a school system to accept and adopt a new technology. Many changes and adaptations have to be made for a smooth adoption. For example, teachers need time to explore new technologies, and training might be necessary to get them started. New policies and regulations are often needed to deal with management issues such as safety, security, and appropriate technology use. Fourth, a fundamental issue is how to use handheld computers in meaningful ways to facilitate teaching and learning. Research on this aspect has been scarce, and this calls for efforts from policy makers, educators, researchers, and classroom teachers.

A trend of technology integration in schools is the use of multiple computing devices at the same time. In fact, there are hundreds of technology options for schools to choose and use.[8] The ecosystem of learning technologies has become incredibly crowded (Dede, 1996), with a growing number of devices. Some schools are being remodeled or rebuilt to better accommodate emerging technologies and better facilitate mobile learning. For example, Maryland's Charles County Public Schools spent $50 million to build a new high school that would "integrate advanced technology into its fabric and culture" (Cevenini, 2006). This school has a range of technologies, including an IP network that supports multimedia data, a wireless network for the whole school, IP phones, wireless IP phones, LCD projectors, desktop PCs, laptop PCs, and Tablet PCs (Cevenini, 2006). In Philadelphia, $63 million was spent on building the "school of the future" in conjunction with Microsoft Corp. to provide high-tech tools to support hands-on instruction and collaborative project-based learning (Murray, 2005). Similar schools have been built in other countries such as the United

Kingdom[9] and Australia to better host the technology ecology. It will be increasingly common for teachers and students to use different ubiquitous computing devices simultaneously.

Like different species in the same ecosystem competing for resources, the different computing devices can also compete with or complement each other (Zhao & Frank, 2003). How to take full advantage of the available technologies and to use different computing devices in complementary ways will be another issue that needs close attention.

Professor Bell is convinced that in the near future it will be possible to see by telegraph, so that a couple conversing by telephone can at the same time see each other's faces. Extending the idea, photographs may yet be transmitted by electricity, and if photographs, why not landscape views? Then the stay-at-home can have the whole world brought before his eyes in a panorama without moving from his chair.

The Industrial Development of Electricity, **1894, quoted in Facer, J. Furlong, R. Furlong, and Sutherland, (2003, p. 223)**

Introduction

1997

A book-lined study, late at night. The house, to which the study is an extension, is quiet. Everyone except for a middle-aged man is asleep. He is sitting at a desk, peering at a computer screen, occasionally moving a mouse which runs on a mat on the desktop and clicking the button which sits on the back of this electronic rodent. On the screen a picture appears, built up in stages, line by line. First it looks like a set of colored bars. Then some more are added, and gradually the image becomes sharper and clearer until a color picture appears in the centre of the screen. It is a cityscape. In the foreground are honey-colored buildings and a solitary high-rise block; in the distance is the sea. Away to the left is a suspension bridge, wreathed in fog. It is in fact a panoramic view of San Francisco, snapped by a camera fixed on the roof of the Fairmont Hotel on Nob Hill. The bridge is the

Golden Gate. Underneath the photograph is some text explaining that it was taken three minutes ago [and] will be updated every five minutes. The man sits there patiently and waits, and in a few minutes the image flickers briefly and is indeed rebuilt before his eyes. Nothing much has changed, except that the camera has moved slightly. It has begun its slow pan rightwards, towards the Bay Bridge. And as the picture builds the solitary man smiles quietly, for to him this is a kind of miracle ...

Naughton (1999, p. 1)

One-to-one computing not only influences how children learn and interact in schools, but also changes the dynamics at home, which is a critical part of children's development and exerts further impact on learning in schools. As we noted in chapter 2, the equity argument for one-to-one computing access in schools encompasses the notion that it will lead to more equal access in homes (Hoppe, 2006; Philpott, 2006; Zucker & McGhee, 2005). For example, the evaluation of Henrico County School District's laptops initiative documented that it had resulted in "greater access to resources and information for more students and families" (Zucker & McGhee). As an Apple-sponsored review of one-to-one computing noted in relation to home computer use, "Being able to take computers home further expands students' access, facilitates students keeping their work organized, and makes the computer a more 'personal' device" (Zucker & McGhee, p. 1).

However, increased one-to-one access to computers by children and adolescents at home not only impacts homework and school-related learning, but also increases the likelihood that children will use computers for gaming, online networking, and other play or social activities. Over the course of the 2 years (2005 and 2006) in which we wrote this book, there was a revolution in young people's (mainly adolescents') use of online social networking sites such as Bebo, MySpace.com, and Facebook, among others (Lenhart & Madden, 2007; Valkenburg, 2006). This revolution makes clear the fast-changing nature of students' digital lives and the ways in which children's and adolescents' increased access to one-to-one computing serves their personal and social aspirations as well as schools' educational goals. However, we situate our discussion of the fast-changing nature of technology use in homes by children and adolescents in a historical context by looking at the evolution of young people's technology use at home.

Thus, in writing this chapter on computers at home and the impact of one-to-one computing on children's digital experiences in this context, we will also be considering it in the context of a long history of technologies entering the home, not the least of which is the television (Liebert & Sprafkin, 1988; Wartella & Jennings, 2000; Winn, 2002). Research on television use in homes has been extensive over the last four decades and it still is the dominant type of technology use by children and adolescents despite the undeniable appeal of computers (Rideout, Fochr, & Roberts, 2005; Rideout, Ula, Foehr, Roberts, & Brodie, 1999).

To understand fully the penetration of one-to-one computing into students' lives, this chapter examines the mutual influences between one-to-one computing and student activities at home, their relationships and interactions with parents and siblings, and other aspects of the home environment. Drawing on data from the United States and England in particular, we address a number of aspects of the computing environment in the home:

- the way in which the home has increasingly become a hub for multimedia opportunities for children and adolescents (including patterns of computer usage by children and their parents)
- the different meanings of the desktop computer or, more recently, the laptop in the home environment
- the "digital gap" and home computing environment in the United States (mainly)
- schools' impact on digital literacy in the home
- parental involvement in children's school learning through technology
- parents' concerns over children's computing at home
- parental supervision of computing at home and out of school

The move toward one-to-one computing points to the way in which unprecedented access to computing promises more equal access to computing—not only in schools, but also in homes where digital tools are increasingly ubiquitous (Rideout et al., 2005). Laptop projects are likely to become an increasingly significant bridge between home and school. Unlike the mobile phone, another portable and ubiquitous digital tool whose use in homes has grown organically, school-initiated laptop use at home has had to adopt proactive strategies to enhance the latter's use at home. Thus, attention to the home component of school-based laptop initiatives forces us to look at the readiness of both home and school to support laptop use to further educators' and families' educational goals. Compared to the now burgeoning literature on the school components of laptop initiatives, the home component has not been the focus of a comparable degree of attention (e.g., Apple, 2005; Conway, 2005; Daly, 2006; Zucker & McGhee, 2005). Nevertheless, we highlight some insights from studies of the home component of school-based laptop initiatives, as well as the use of other digital tools (e.g., Penuel et al., 2002; Rideout et al.).

The arrival of the net generation

New homes come complete with special nooks for oversized TV screens and home entertainment centers, while new cars come with personal TV screens in the back of each seat. The amount of media a person used to consume in a month can be downloaded in minutes and carried in a device the size of a lipstick tube. Today we get movies on cell phones, TVs in cars, and radio through the Internet. Media technologies themselves are morphing and merging, forming an ever-expanding presence throughout our daily environment.

Rideout et al. (2005, p. 4)

In order to set a context for this chapter, we draw upon a 1999 survey that demonstrated the closeness of the relationship between computers and users in the United States—whether the users were adults or children. As such, it is reasonable to conclude that homes are fertile ground for one-to-one computing, given the appeal of computers to both children and adults. The comparison between adults' and children's responses reveals some important and consequential differences in terms of how home computer use may be construed differently by adults and children. As one-to-one computing promises to increase the degree of engagement with computers in homes as well as schools, understanding children's and adults' overlapping and at times contrasting experiences with and attitudes toward computers is important, in terms of how computer access and the meaning of this access are socially constructed in homes.

The joint NPR, Kaiser, and Kennedy School of Government 1999 nationally representative survey found that the vast majority of people in the United States thought that computers had improved their quality of life. The authors of the report concluded that a "love affair" exists between computers and their users. At the same time, a separate nationally representative sample survey of 625 children and adolescents aged 10–17 revealed that they were even more positive than their parents. The survey authors concluded that "enthusiasm for computers and the Internet runs wide and deep, across all incomes, all regions of the country, all races, all political ideologies, and most age groups" (http://www.npr.org/programs/specials/poll/technology/).

For example, children reported that they were more enthusiastic about and comfortable with computers than adults were. Whereas 85% of children reported that they were keeping up with computers and only 14% thought they were being left behind, adults were evenly split between the 49% who said they were keeping up and the 49% who said they were being left behind. It is indicative of the perceived social necessity of computing in today's child and adolescent peer groups that children without computers were more concerned that they were missing something than adults were. Furthermore, children and adolescents who did not have a computer at home were far more likely than such adults (42 vs. 23%) to think that not having a computer at home was a problem. The latter figures suggest that children and adolescents today see computers as an essential aspect of childhood and adolescence and that not having access leads to perceptions of social exclusion.

Taken together, the preceding NPR/Kaiser/Kennedy School of Government report and U.S. Census Bureau 2005 figures documenting children's computer access at home illustrate two important points: Access has grown steadily over the last 20 years, and a large proportion of children from 3 to 17 years of age of have access to computers at home (see Table 6.1). For example, in 2003, 70% of 3- to 5-year-

Table 6.1   Home computer access and home internet use, 1984–2003[a]

Year	Home computer access (%)	Home Internet access (%)
1984	15.3	—
1989	24.2	—
1993	—	—
1997	49.7	21.7
2000	65.0	30.4
2001	70.4	40.1
2003	75.5	42.0

[a] Children aged 3–17.
Source: U.S. Census Bureau, 2005.

olds had access, compared with 79% among 15- to 17-year-olds. Home Internet use increases substantially with age. However, only 15% of children aged 3–5 used the Internet at home in 2003, compared with 65% of 15- to 17-year-olds.

More recent statistics from the Pew Internet and American Life Project, based on a nationally representative survey of 12- to 17-year-olds in the United States, point to how teenagers are leading the charge toward a mobile and fully wired nation. The Pew report (Lenhart, Madden, & Hitlin, 2005) documented a situation in which teenagers in 2004 were significantly more involved in the use of portable and wireless computing even than their peers from the comparable 2000 survey. For example, in 2004, 87% of teenagers aged 12–17 reported they used the Internet, compared to 73% in 2000. In contrast, 66% of adults used the Internet compared to 56% in 2000.

The frequency of Internet usage was also a good indicator of changing trends among teenagers, with 51% reporting that they go online on a daily basis, up from 42% in 2000. The Pew report portrays a cohort of teenagers significantly more involved with computers than the equivalent teenagers merely 4 years earlier. The Pew report documented teenagers who often used more than one digital device, with 84% reporting that they owned at least one personal media device (i.e., a desktop or laptop computer, a cell phone, or a personal digital assistant [PDA]). Almost half (44%) said they had two or more devices, 12% had three, and 2% reported that they had all four devices. Only one in six (16%) reported that they did not have any of the four devices.

In summary, the preceding "whole population" perspective illustrates a number of important points in understanding the nature of one-to-one computing in homes:

- Both adults and children like computers, but children are more likely to perceive them as necessary.
- One-to-one computing is not a future aspiration, but rather a reality today for many children and adolescents, given the proliferation of digital tools from desktop and laptop computers to PDAs and cell phones (Rideout et al., 2005).
- Home is a significant location in understanding children and their access to, use of, and understanding of computers. For most children, home provides access to many different digital tools as well as connectivity (conduit access, see chapter 2).

Table 6.2   Computer access and use at home[a]

Country	Percent with home access	Percent using Internet at home (of those with computers at home)	Percent saying, "I am not allowed to give out personal information online"	Percent saying, "I have rules about how much time I spend online"
Denmark	77	81	49	26
Iceland	85	85	39	28
Ireland	80	80	70	49
Norway	73	80	47	34
Sweden	87	87	60	32

[a] Children aged 9–16.
Source: NCTE/SAFT, 2006.

- The number of children and adolescents with both computer access and online access at home has increased significantly over the last 4 years.
- While this "whole population" perspective is illuminating in conveying overall growth in access to computers and online access at home as a very significant social trend, it also masks important social class and ethnic group differences in each of these categories. We will return to this point later in the chapter.
- We tend to agree with the Pew report (Rideout et al., 2005) conclusion that today's children and adolescents are leading the advance into a more wired and mobile digital world—that is, they are willing, wired, and mobile. Young people's access to more than one digital device and increasing opportunities for connecting devices wirelessly or via external devices such as USB memory sticks are creating interdevice connectivity possibilities that, we think, will increasingly create new synergies in the use of digital tools.

A final word: In this section we concentrated on statistics from the United States, but if we had chosen any number of other developed countries the picture would have been very similar. For example, a recent report on 9- to 16-year-olds' access to and use of computers at home in five European countries (Denmark, Sweden, Norway, Iceland, and Ireland) documented figures very similar to those in the United States (see Table 6.2). The survey sought to document children's access to and use of the Internet and identify potential risks and dangers for children in cyberspace (NCTE/SAFT, 2006).

The persistence of old technologies and the rise of new ones

New technologies—especially those with more intense opportunities for engagement in one-to-one computing where technologies are net-worked—move the home beyond merely being a site for children to use and receive various technologies to becoming multimedia hubs that give children the capacity to communicate digitally in text, audio, and image formats with others in distant locations. Children in devel-oped countries, in particular, live in a world in which media and vari-ous digital technologies constitute an increasingly significant aspect of their lives (Lenhart & Madden, 2007; Papert, 1993; Rideout et al., 1999, 2005; Villani, 2001). Rideout and colleagues' U.S. study, based on a national sample of 2,032 students in grades 3–12 (8–18 years old), noted that "two thirds (68%) have a TV in their bedroom, half have a VCR/DVD player (54%) and a video game player (49%), and nearly one third (31%) have a computer in their room" (p. 5).

However, one particular medium, television, has been a very sig-nificant aspect of children's lives for the last few decades, and research on children's television provides ample evidence that 10- to 12-year-old children watch television for about as much time as they spend in school, with studies indicating weekly viewing of between 22 and 30 hours (Annenberg Public Policy Center, 2000; Liebert & Sprafkin, 1988; Rideout et al., 2000). Neither younger children nor adolescents, who also watch television for significant numbers of hours, watch as much as 10- to 12-year-olds (Liebert & Sprafkin; Rideout et al.).

These long-standing television-viewing patterns alert us to the pos-sible patterns in children's use of all kinds of media including digital technologies. For example, Rideout and colleagues' 1999 study of chil-dren's media habits in the United States "identified a number of trends and usage patterns that indicate children today live in an increasingly media-saturated environment." Drawing on a nationally representative sample of 3,155 children aged 2–18 and using a combination of ques-tionnaire results and week-long media-use diaries kept by 621 of the children surveyed, the researchers concluded that American children, even very young children,

> … spend the equivalent of a full-time work week using vari-ous types of media: 5 hours a day, 7 days a week, for a total of more than 38 hours a week (not including media used in the classroom or for homework purposes). Children aged 8 years and older spend even more time with media, averaging 6 hours a day.

Children between the ages of 2 and 7 spend an average of 3 hours a day using media.

They also noted that children typically had access to a number of media outlets at home, including three television sets, three tape players, three radios, two VCRs, two CD players, one video game console, and one personal computer. However, in terms of computer use, they made an intriguing observation:

> Contrary to expectations, although access to computer is widespread among children, they spend a relatively short amount of time with computers each day. The average child will spend little more than half an hour a day (34 minutes) on a computer. Fewer than one tenth (9%) of children spend more than an hour a day using computers "for fun," and only 3% spend more than one hour a day on the Internet. More traditional forms of media, especially television, dominate children's media consumption. (p. 39)

These researchers noted a growing "bedroom culture" among children, where bedrooms seem to be rapidly becoming what they called "media central," providing most children the opportunity to consume many different kinds of media in the relative privacy of their own bedrooms. They noted that children's bedroom media included radios (70%), tape players (64%), televisions (53%), CD players (51%), video game consoles (33%), VCRs (29%), and computers (16%). In relation to children 8 years and older, 65% had a television and 21% had a computer in the bedroom. Interestingly, in terms of the appeal of computers, they found that, when asked to pick just one form of media they would prefer to have with them on a desert island, children 8 years and older were three times more likely to choose computers (33%) than television (13%). Furthermore, even among 2- to 7-year-old children, 23% chose computers, 29% selected television, and 23% picked video games.

Rideout et al. (1999) also identified other important disparities in children's media consumption. First, they found that children who lived in or went to schools in "lower income communities" tended to spend more time with most types of media than children from higher income neighborhoods or schools. However, one critical exception was computer use. That is, children from lower income communities "were significantly less likely to use computers—only 29% used computers on a typical day, compared to 50% of kids living in wealthier communities." Based on these disparities along socioeconomic lines,

Rideout and colleagues claim that "schools are helping to equalize access to computers" since children from lower income communities were just as likely as higher income children to use computers at school (32 vs. 30%).

With such pervasive domestic media consumption by children and adolescents, some researchers have begun to ask what effect this might be having on children. In her historical review of research into the impact of media on children and adolescents published during the 1990s, Villani (2001) examines how concerns and theories involving more traditional forms of mass media translated to new media formats such as the Internet and video games. She notes that a large proportion of the research conducted up to the 1990s focused on behavioral and ideological "media effects"; that is, research focused on how children's behaviors and value systems were being shaped by the media.

The titles of two widely cited 1980s books on the impact of television on children, *The Early Window* and *The Plug-In Drug*, capture the two aspects of this focus: Should television be viewed as an early learning window or an addictive "plug-in" drug? However, she claims that, since 1990, the approach of most research has shifted, emphasizing instead media content analysis and discovering children's viewing and media usage patterns (e.g., the studies by Facer et al. and Rideout et al. noted earlier). Villani (2001) goes on to conclude, in an almost alarmist fashion reminiscent of concerns raised in the 1970s and 1980s about the impact of television on children's levels of aggression, that prevalent new media usage habits are reason enough for parents and researchers to be concerned about the media, given its "tremendous capacity to teach" and the possibility of its fostering "social isolation" and even a distorted world view with a potential increase in high-risk behaviors.

Concerns about the impact of television, computer, or one-to-one access to various digital tools on children often miss the more subtle and varied meanings that specific cultural tools take on in particular contexts. While the desktop or laptop computer has the potential to be many different tools, especially if we take into account the vast range of software now available, singular statements lamenting the sad impact of computing on children may ignore the rich variety of meanings accorded computers in different families. We now turn to this theme, drawing in particular on a large-scale study, *ScreenPlay: Children and Computing in the Home,* undertaken in England and Wales between 1998 and 2000 (Facer et al., 2003).

The computer as children's machine, as interloper, and at the heart of the family

> Computers did not have a "fixed" and "universal" identity within different families.
>
> **(Facer et al., 2003, p. 62)**

As we noted earlier, the domestication of the computer is one of the significant developments in computer use over the last decade. However, this general observation misses important differences in the meaning of home computer use by children and adolescents. In order to explore these differences in meaning, we draw upon an ethnographic study undertaken in England and Wales. Based on their analysis of the different ways computers were used in the home, Facer et al. (2003) used three different metaphors to distinguish children's and adolescents' computer use at home: "the children's machine," "computer as interloper," and "computer at the heart of the family."

In families (most families) where the computer was seen as the children's machine, the computer was an intrinsic part of family life in terms of both leisure and work. The computer was an embodiment of parents' aspirations for their children's future welfare. This attitude was most prevalent in homes where parents had the least experience of computers in their own workplace. In families where the idea of the computer as interloper dominated, the computer was seen as in conflict with dominant family values, as was often manifested in arguments between parents and children or among siblings about its role. In families where the computer was at the heart of the home, it played a very significant role in many different ways, encompassing work and leisure, and was the source of new relationships between family members. In contrast to the "children's machine"-oriented families, in those where the computer was at the heart of the family there was more involvement by all family members in collaborative activities and sharing expertise.

Critical of studies that rely on headline-catching statistics to convey the way in which today's children are engaging with digital technologies, Facer et al. "followed the child"; in addition to a survey of 855 children and adolescents from 8 to 17 years of age, they undertook detailed case studies of 18 children over a 2-year period (1998–2000). They also conducted focus group interviews with 48 young people who described themselves as low users of computers. In following the child, the researchers sought to understand the meaning of computers to children within their families and the construction of their individual and family histories in relation to technology.

England and Wales, like the United States and many other developed countries, have experienced a rapid rise in the percentage of homes with computers as well as the percentage of homes with Internet access. For example, Facer et al. noted that various studies from 1998 to 2000 indicated that between 69 and 88% of young people had computer access at home. They also reported a rapid increase in Internet connectivity even over the relatively short life span of their study: from 25% in 1998 (in an initial scoping survey in their own research) to figures of 64 and 75% in two surveys undertaken in 2000 (Pathfinder) and 2001 (InterActive), respectively.

Home computing environment in the United States

Research in many countries has documented unequivocally that a "digital gap" in the home computer access and online access patterns exists between children from low socioeconomic status (SES), low status ethnic groups, and middle/high SES groups and high status minority groups (Beamish, 1999; Becker, 2000; Facer et al., 2003; Koss, 2001; U.S. Census Bureau, 2005; Warschauer, 2003). The advent of one-to-one computing, like each wave of innovation in digital technologies, changes definitions of the digital divide. Research on the dynamics of the digital divide, if we frame it in terms of access to devices and access to conduits (or connectivity; see chapter 2) reveals, as we have noted, a clear picture of significant differences between low- and high-status groups along lines of class and ethnicity.

However, the digital gap can be analyzed at yet another level and that is in terms of meaningful engagement in social practices. As we noted in chapter 2 (Warschauer, 2003), access to devices is measured by some index of density in particular contexts. In the case of homes, it might be the percentage of homes in a country or state with a PC. Alternatively, it might be the number of PCs per home, which will yield a different picture of the digital divide. A focus on conduits will address the number and nature of connectivity to the Internet. This, too, provides an important indicator of the digital divide. Rideout et al. (1999) found that most children (69%) have a home computer and that nearly half (45%) had Internet access, with10% having Internet access from their bedroom. However, these researchers also noted that children from lower income communities "were significantly less likely to use computers—only 29% used computers on a typical day, compared to 50% of kids living in wealthier communities."

It is only by paying attention to the degree of desktop or laptop computer use that some of the more subtle digital divide gaps appear. Furthermore, Facer and colleagues' (2003) study provides significant evidence that ethnographic studies may be essential to understanding the differences in access to devices, access to conduits, and access to meaningful computer practices—not only among social classes and ethnic groups, but also within families in gaps in access to the computer as a resource. Thus, for example, they noted that expertise and gender played a role within families in facilitating or inhibiting children's engagement with computers at home.

Parental involvement through technology

> Schools are the crossroads of yesterday's traditions, today's demographics and tomorrow's technologies.
>
> **Epstein (1985, p. 1)**

> As families' access to advanced computer and telecommunications technologies has increased, new opportunities to forge home–school connections supported by new and advanced technologies have become possible. Many new programs have been implemented in recent years that make available to students desktop computers for use at home. Other programs provide laptop computers that can be taken back and forth between home and school.
>
> **Penuel et al. (2002, p. iii)**

The potential of technology to enhance parents' understanding of what their children are learning in school is a relatively undeveloped aspect of research. In the particular case of computing, there have been very few rigorous studies that examine the contribution of technology programs to student learning and parent–school communication. However, Penuel et al. (2002) undertook a meta-analysis, selecting 19 of 98 studies on the role of technology and home–school connections.

Based on their meta-analysis they concluded that technology integration programs designed to improve home–school connections typically result in:

- A modest increase in student reading ability. However, this improvement is smaller than in other sorts of intervention programs, such as class size reduction, computer-assisted instruction, tutoring, individualization, increased homework, coaching for achievement tests, and parental involvement programs.
- A substantial increase in student writing ability. This improvement is comparable in size to that typically found in integration programs that stress computer-assisted instruction, and is larger than the effect sizes associated with intervention programs that stress class size reduction, individualization, or coaching for achievement tests.
- A modest increase in student mathematics ability. This improvement is comparable in size to that typically found in integration programs that stress coaching for achievement tests, and is larger than that typically found in class size reduction and individualization intervention programs.
- Communications between parents and school are significantly improved.

These findings suggest a moderate but not transformative role for technology in enhancing student learning and a more significant role in terms of the potential of technology to increase and enhance communication between home and school. This latter finding is especially important in the context of one-to-one computing as it may facilitate greater communication between schools and parents.

From school to home: supporting laptop use at home

Facer and colleagues' study (2003) is helpful in understanding how computers that already exist in the home are used by children and parents. However, many one-to-one laptop projects seek to improve students' computer access at home by providing either more access where it already exists or better access where it does not exist at all. Studies of the Laptop Initiative for Students with Dyslexia and other Reading and Writing Difficulties (LISD) in Ireland provide a helpful framework for understanding the dynamics of laptop deployment in schools and how this may or may not facilitate their migration to home use (Conway, 2005; Daly, 2006). The emergence of three types of laptop deployment that typically coexisted in the schools reminds us how educators and policy makers' aspirations that laptops become "personal" learning tools for students are mediated by meanings and practices that emerge in specific school and system contexts. Schools in the study deployed laptops in one or more of the following ways: fixed, floating, and fostered.

Pedagogical dimension	Mobility	Personal tool	Technology integration
Deployment model			
Fixed	Low	Low to moderate	Low
Floating	Moderate	Moderate to high	Low to moderate
Fostered	High	High	High

Source: Conway, 2005, p. xiii.

The fixed model occurred when laptops were allocated to one "fixed" physical setting in the school. For example, in some schools students only used laptops when they visited the "laptop room" during single or double lesson periods 2 or more days per week. This fixed model resulted in low levels of laptop mobility, decreased the likelihood of students' chances of using laptops as personal learning tools, and short circuited efforts to use laptops as a means of fostering high levels of technology integration along lines outlined in Sandholtz, Ringstaff, and Dwyer's ACOT study (1997).

In the floating model, students were able to use the laptops flexibly in different locations in the school. This meant that students carried laptops around with them or the laptops were transported via a laptop trolley in order to give students easy access in different lessons. The key difference between the fixed and floating models was that students had more sustained access in different locations to laptops. The floating model resulted in greater laptop mobility, increased the likelihood that the laptops were used as personal learning tools, and improved chances that technology integration was enhanced. Use of laptops at home by students was not a feature of either the fixed or floating models.

On the other hand, the fostered model linked a laptop to a student and typically facilitated students' laptop use both at home and at school, although in some instances this came to mean the laptop was designated for a student's use either only in school or only at home. The key insights from the range of laptop deployment models used in the LISD project was that schools mediate how laptops are deployed, that while different deployment models may coexist, one typically dominated (i.e., the fixed model), and, most importantly, that the extent to which laptops were ever used by students at home depended on the whether the fostered model was actively supported in each school.

How was laptop use at home actively supported in various one-to-one initiatives? When laptop use at home actually occurred in a laptop project it typically involved schools in setting expectations and rules about appropriate laptop use to foster a learning-oriented context for their use at home (Bonifaz & Zucker, 2004; Daly, 2006; Zucker & McGhee, 2005). As such, the optimal use of laptops for learning by students at home demands considerable planning and support by schools as well as parental involvement. Among the many essential and common features of successful laptop use at home that have emerged as important across these varied evaluation studies were purchase of adequate insurance cover (Bonifaz & Zucker), attention to safety and security of laptops (Bonifaz & Zucker; Daly), development of rules for laptop use (Daly), setting up filters to limit accidental or deliberate misuse of laptops at home (Bonifaz & Zucker), assigning homework that necessitated laptop use at home (Bonifaz & Zucker; Daly; Zucker & McGhee), supporting or subsidizing broadband access at home (Bonifaz & Zucker; Zucker & McGhee), and providing information and/or training for parents (Bonifaz & Zucker).

Even with these supportive dimensions in place, parents often find they have concerns about their children's computer use, especially in the context of one-to-one use since it may increase the likelihood that children engage in inappropriate computer use. For example, the meteoric increase in online social networking has resulted in considerable concern by parents and educators about the amount of time expended online by adolescents as well as the nature of their online activities. In this regard, concerns about the impact of online social networking are not unlike concerns expressed a generation ago about the addictive and aggression-inducing potential of television (Winn, 2002)

Parent's concerns over children using computers at home

Across the world, children have entered into a passionate and enduring love affair with the computer. What they do with computers is as varied as their activities. The greatest amount of time is devoted to playing games, with the result that names like Nintendo have become household words. They use computers to write, to draw, to communicate, and to obtain information. Some use computers as a means to establish social ties, while others use them to isolate themselves. In many cases their zeal has such force that it brings the word addiction to the minds of concerned parents.

Seymour Papert (1993, p. ix)

Caution: children at play on information highway; access to adult networks holds hazards

Genevieve Kazdin, a self-appointed crossing guard on the information highway, remembered the day last September when she found an 8-year-old girl attempting computer conversations with a group of transvestites. Seemingly safe at home, the child was playing with her favorite $2,000 toy, using her computer and modem to make new friends through a service called America Online. The name of the electronic discussion group the girl had discovered was called, confusingly enough, "TV chat"—the TV being shorthand for transvestite. Kazdin said the girl had read it differently: "She was thinking in all innocence, 'We're going to talk about Barney.'" Kazdin recognized the girl's "screen name" because the Massachusetts grandmother helps run America Online's programs for children. Kazdin chatted with her little friend via keyboard, gently steering her to a more appropriate part of the service—and preventing one of a growing number of daily culture shocks as users wander into rowdy neighborhoods found in the new online community.

John Schwartz, *Washington Post*, November 28, 1993, p. 1

This 1993 front-page story from the *Washington Post* reminds us that the Internet and online environments have now been around for a considerable length of time. Today's teenagers and children will have grown up in an era in which the Internet and cyberspace were ordinary everyday aspects of life rather than new-fangled inventions. A 2003 report of a survey in England, Scotland, and Wales reported that 93% of parents were confident in their supervision and care of their children's Internet use. However, this still left thousands of children in potentially unsafe situations where their parents felt less than adequately prepared to set rules for online life in their homes.

In the United States in February 2006, i-Safe, an Internet safety non-profit organization, announced a partnership with Microsoft to provide free on-demand Internet safety resources for parents and educators. "A survey by our National Assessment Center revealed that more than half of parents feel their ability to monitor and shelter their children from inappropriate material is limited," said Teri Schroeder, founder and CEO of i-Safe. The i-Safe program encompasses six online education video modules, which address personal safety, cybercommunity issues, cyberpredator identification, cybersecurity, intellectual property, and community outreach.

In discussing parental concerns about computers, the continued increase in the amount and pervasiveness of online activity by children and adolescents is often of particular concern. For example, the 2006 NCTE/SAFT survey of 9- to 16-year-olds in five European countries, cited earlier in this chapter, documented that between 25 and 35% of children surveyed had received undesirable sexual communications while using the Internet. The survey also questioned the children about mobile phone usage and reported that 26% of mobile phone owners had received a message that threatened or frightened them; 41% of children who received such messages knew the person who sent the message; and 42% reported no rules from home for using mobile phones.

Consideration of such occurrences by parents and educators is often cause for real concern even when they know that children's motivation for engaging in online activities and rapid adoption of digital technologies is primarily driven by three motives: entertainment, personal connections, and empowerment (Izenberg & Lieberman, 1998). Izenberg and Lieberman also noted that some research hypothesizes that online encounters can offer children a refuge from real-world bullying by providing support from others based on interests and personality rather than exterior appearance and social norms. Facer et al. (2003), for example, claim that despite the potential educational benefits of online activities undertaken from the safety of the home, the potential dangers for children on the Internet make it more like a dangerous street than a safe haven and sanctuary.

In the light of the range of concerns being raised by parents, various agencies—from Internet service providers (ISPs) to governments, education ministries, and educational technology agencies—have been developing materials to foster safer online experiences and behavior for children. With significant overlap in approach and materials given the common problems being addressed, many of these include a parental guide and online safety handbooks replete with examples of positive approaches to creating safe cyberspace experiences for children and adolescents. High on the agenda in these guidelines is the role parents can play in choosing and using blocking and filtering software, as well as helping their children to become more critical consumers of information available on the Internet. The NCTE/SAFT report noted that up to half of the children and adolescents surveyed believed the Internet provided accurate and truthful information.

Children as victims and perpetrators

For a small minority of children and adolescents, the virtual world poses very significant personal dangers in the form of bullying or being the victim of predatory sexual behavior (Erooga & Masson, 1999; Quayle, 2004; Taylor & Quayle, 2003). For victims, what appear like innocent conversations in harmless chat rooms may sometimes be the early stages of grooming by adult males (typically) who adopt a persona or screen name and create on online identity in order to develop a relationship with a child. The international nature of these online activities has necessitated the involvement of police forces and intelligence agencies across borders to identify and apprehend perpetrators as well as identify and provide support for child victims of online sexual behavior (Taylor & Quayle, 2003).

Supervising computing at home and out of school: suggestions

Each wave of technological innovation has brought its share of concerns as parents and educators have considered not only its educational potential but also the damage it may cause to children (Wartella & Jennings, 2000). In the case of television, considerable research was undertaken over a 20-year period to assess whether or how television might foster aggressive behavior among children and adolescents (Liebert & Sprakfin, 1988). Parents in the United States are still more concerned about television use than computer use, despite changing media consumption patterns at home (Annenberg Public Policy Center, 2000). In relation to computer use at home, parental concerns center on a number of issues: time spent using computers at the expense of other activities, accessing pornography, giving out personal information, meeting social or sexual predators in cyberspace, and taking up an invitation offered online for a face-to-face meeting.

As we noted earlier, there is now an extensive array of material available for the promotion of Internet safety, all of which becomes even more pertinent in a one-to-one computing era. In Ireland, for example, Webwise provides parents, teachers, and children with educational resources, advice, and information about potential dangers on the Internet (see www.ncte.ie). Webwise shares strategies, information, and resources with similar groups across Europe through the European Commission's Insafe network. The Webwise site (http://www.webwise.ie) provides extensive material for parents, teachers, and children under four key headings—*surfwise, chatwise, sharewise,* and *gamewise*—as a way of identifying safe practices for children and adolescents in their everyday experiences of online life.

Online friendship networks: parents' and adolescents' online play at home

> *TIME Magazine* Person of the Year 2006: YOU. Yes, you. You control the Information Age. Welcome to your world.
>
> *Time Magazine,* **front cover,**
> **December 25, 2006–January 1, 2007**

In the past five years, such sites have rocketed from a niche activity into a phenomenon that engages tens of millions of Internet users. The explosive growth in the popularity of these sites has generated concerns among some parents, school officials, and government leaders about the potential risks posed to young people when personal information is made available in such a public setting.

Lenhart and Madden (2007, p. 1)

The choice of "YOU"—the ordinary digital citizen, digital citizen Joe or Josephine—as *Time Magazine*'s person of the year for 2006 recognizes a significant social revolution occurring in our midst. Adolescents, especially, play an important role in the burgeoning use of friendship networking (e.g., MySpace.Com, Bebo, and Friendster) sites and expression sites such as YouTube. In framing a study of Dutch adolescents' use of friendship sites, Valkenbourg, Peters, and Schouten (2006) spoke of the "worldwide proliferation of such sites" as a very significant context within which to understand Internet communication—both its peril and appeal.

Increased one-to-one computer access is likely to widen and deepen adolescents' engagement with online networking and online digital expression sites. Rather than only conversing or playing on the street, adolescents are now also turning to online modes of communication that they typically access at home. The contemporary online social networking revolution presents new challenges for parents and educators in setting and managing boundaries for children's and adolescents' screen lives (Lenhart & Madden, 2007; Valkenbourg et al., 2006). As such, consideration of how best to create safe online social space for young users of these sites has put pressure on site owners to engage in some surveillance and proactive management. For example, Bebo appointed a safety officer in 2006 whose professional background was in studying the dark side of the Internet—namely, child pornography. Such is the seriousness with which new areas of professional expertise are emerging to support ICT professionals, parents, and educators in ensuring that young people's online social lives exert a positive influence on their personal and social development.

Given that access to online friend networking sites typically occurs out of school rather than in it—that is, at home or another nonschool site, parents in particular are being challenged to set appropriate boundaries and ground rules for this aspect of their children's and adolescents' lives. However, one of the challenges for educators, researchers, and families is the need to understand the actual nature of young people's motives and experiences as they engage in online social life. Recent large-scale studies in the Netherlands (Valkenbourg, Schouten & Peter, 2005; 2006) on the dynamics of adolescents' social behavior on friend networking provide some important new evidence that might allay parents' fears about what for many may be a relatively hidden aspect of their children's lives. Based on their study ($n = 609$) of adolescents' online identity experimentation, they note that "the most important motive for such experiments was self exploration (to investigate how others react), followed by social compensation (to overcome shyness) and social facilitation (to facilitate relationship formation)" (2005, p. 383). As such, while parents may be concerned and are increasingly likely to be so in the future as one-to-one computing proliferates, it appears that the motives driving adolescents' online social lives may not be that different from those for their offline social lives.

Conclusion

... children are the epicenter of the information revolution, ground zero of the digital world ...

Katz, *The Right Kind of Kids in the Digital Age* (1997), quoted in Facer et al. (2003)

Machines that fit the human environment instead of forcing humans to enter theirs will make computing as refreshing as a walk in the woods.

Mark Weiser (1991)

We started this chapter with two rather long quotations, each conveying a new vision of how the outside world can enter into the relative privacy of the home. The first, from the perspective of 1894, was truly visionary, recounting Alexander Graham Bell's dreams about the potential of the telegraph to transmit images into people's homes. The second, written by Naughton just over 100 years later, in 1997, is testament to the fulfillment of Bell's dream. In terms of our focus on children and computing at home, the two quotations highlight the way in which technology has changed the boundaries between home and the wider world. In this chapter we have looked at some aspects of these changes, focusing on such issues as children's media consumption at home, the domestication of the computer, the digital divide and computer access and use at home, and parents' concerns about and supervision of children's computer use at home.

The Children's Machine is the title of one of Seymour Papert's books on children and computing. While computers have clearly affected children's home lives in significant ways, we tend to disagree with Papert that the computer is *the* children's machine—at least not yet. Older technologies still have their appeal, and computers have yet to overtake television in terms of time spent with them, even if computers and digital tools are inherently appealing to children. Two other points are important here. First, Facer and colleagues' (2003) study examining the meaning of the computer in children's home lives, as well as various media consumption surveys, points to a pattern of mixed media use rather than dominance by computers. Second, the undeniable appeal of computers to children and the recent growth in their access to and use of them at home point toward a future for children that is likely to be increasingly shaped by one-to-one computing. If we doubt the appeal factor, Rideout and colleagues' (1999) finding that, when asked to pick just one form of media they would prefer to have with them on a desert island, children 8 years and older were almost three times more likely to choose computers (33%) than television (13%) provides convincing evidence that, even if children typically spend less time using computers than they do watching television, they value the time more highly.

We have made a number of key points in this chapter:

- The advent of computing and the current push toward one-to-one computing is part of a long history of technologies entering the home (Wartella & Jennings, 2000). The last 5 years have been a period of rapid increase in access to and use of computers in the United States, and research in other developed countries suggests that similar patterns of home computer use by children are occurring widely elsewhere.

- "Whole population" figures tell a story of changing media consumption by children at home. However, a focus on access to devices and conduits in whole population figures leaves out the central issue of the precise role computers play in children's capacity to experience meaningful social practices in the context of their computer use.

- Different attitudes to the computer in homes reveal that there are patterned social practices according to which families manage computers at home. For example, whether a family decides that a computer is a shared social resource in the home or an item for individual private use matters a great deal in terms of how computers play a role in children's lives at home. Thus, as with classrooms and schools, home is a place where computers can take on different meanings In the case of schools, Zhao and Conway (2001) documented different images of computer use in school policy with significant implications for their use in schools—as a stand-alone tool or networked computer, for example. In terms of actual practice in schools, research on a laptop initiative in Ireland for students with literacy difficulties demonstrated how "mobile" laptops came to take on three different meanings that were reflected in laptop deployment in schools: fixed, floating, and fostered (Conway, 2005; Daly, 2006).

- Home-related social exclusion is evident in statistics relating both to access to devices and conduits in many countries. This has increasingly significant implications for those students who do not have home access as the one-to-one computing phenomenon becomes more widespread, as it appears to have done even in the last 4 years according to the comparison between teenagers in 2004 and 2000 in the Pew report.

- The capacity of one-to-one computing to create new synergies between home and school is only beginning to be exploited. The major SRI review documented a dearth of high-quality studies providing evidence of significant effects. However, despite the dearth of evidence, we also note that the rapidly changing design of digital tools is itself creating new possibilities for the use of one-to-one computing at home and between home and school.

- Computers or other digital tools connected to the Internet raise significant questions of safety in homes. Laying down rules for home Internet use necessitates considerable collaboration among home, school, and the wider community. Parents (and teachers) need the requisite knowledge and skills to create safe contexts for children's online experience, as there are serious safety issues and possibly significant dangers for children, (see NCTE/SAFT survey, 2006, comparing Nordic countries and Ireland).

- In a small minority of cases, children's computer use at home poses significant dangers as children may become victims or indeed perpetrators in activities such as bullying or online predatory sexual behavior (Taylor & Quayle, 2003).

- Finally, the changing nature of home computer use, the different meanings attributed to computers at home, the range of potential devices that qualify under the banner "computer," and the range of activities and outcomes (both negative and positive) possible in using computers at home all point to the fact that the very concept of one-to-one computing is a somewhat elusive and slippery construct. As such, it is important to view one-to-one computer use, at home or school, within an ecological perspective. To put this another way, in attempting to understand the changing nature of the computer–child relationship, it is important to see the computer in its historical context, as one of many technologies that has entered the home over the last century. Perhaps the meaning of the computer—that is, understanding the role ascribed to computers by family members—exemplifies how the contemporary possibilities afforded by one-to-one computing, while significant, are not uniform across families. In this sense, home and school are similar in that it is only by paying attention to the meanings and practices of one-to-one computing in these contexts that we can begin to understand its impact.

Evaluating and studying one-to-one computing 7

Great expectations in one-to-one computing policies

An experiment that allows students to tote their own terminals yields better attitudes and academic gains.

Stevenson (1999, p. 18)

Birmingham, along with some other Local Education Authorities around the country, is embarking upon an ambitious and exciting project which will provide all pupils with their own, personal learning device. The goal of one-to-one access to technology is to be realised through an "e-Learning Foundation," a Charitable Trust. AAL is potentially the single most significant educational initiative in decades. The project will transform the way in which children learn in classrooms across the city, and will extend learning beyond the confines of the school day and the school building. It is expected to have a significant effect upon standards of achievement, and will also contribute to the creation of a technologically literate workforce in the next generation. This will enable citizens to take advantage of shifting employment patterns, as new-technology employers are attracted to the city.

Anywhere, Anytime Learning (AAL): Education for the 21st century, Birmingham Grid for Learning, England[1]

The Birmingham Anywhere, Anytime Learning initiative outlines a series of ambitious goals: transformation of children's learning; extending learning beyond the school day; improving standards of academic achievement; contributing to the preparation of a technologically literate workforce; and, finally, enabling citizens to take advantage of shifting economic patterns. The quotation is valuable because it exemplifies the ambitious aspirations associated with many one-to-one laptop initiatives. Thus, the quotation is also typical because the aims of most laptop programs encompass proximal goals—for example, improving academic achievement in core academic areas such literacy and mathematics—as well as a raft of distal goals, such as preparing a more technologically literate workforce ready to meet demands of an ever changing twenty-first century economy and society.

To what extent do laptop programs and other one-to-one computing initiatives live up to these high expectations? To what extent can we say in 2006 that one-to-one initiatives undertaken to date have their espoused proximal and distal goals? Based upon what evidence might we evaluate the numerous aspirations associated with one-to-one programs? How do we evaluate whether so-called twenty-first century skills have been attained using conventional tests of academic achievement? Over what duration ought one-to-one initiatives to be evaluated?

Evaluation: approaches and standards

Evaluation is the process of determining significance or worth. In the context of educational innovations and reforms, this usually involves an analysis and comparison of actual progress versus prior plans, with a view to improving future implementation. Evaluation is now a burgeoning industry as policy makers are increasingly held accountable for the impact of expensive educational and social policies. Given that one-to-one computing initiatives are costly, systematic knowledge about the processes and impact of such initiatives is vital for policy makers rather than relying on a trial and error method (Lipsey & Cordray, 2000).

Whereas evaluation designs in the early days of evaluation studies emphasized experimental designs focused on program outcomes, today's evaluation studies also adopt interpretive designs in order to understand program implementation. As such, in the early days of evaluation, evaluators were, according to Campbell (1971), "methodological servants to the experimenting society"; today, they are increasingly playing a role in program design, planning, and refinement. In the case of one-to-one computing, current evaluation studies are consistent with changes in the wider field of evaluation encompassing both implementation and outcome studies.

This chapter asks what we can learn from evaluation studies of one-to-one computing in order to ask the right questions to envision the design of future initiatives and their evaluations. Thus, our focus is primarily on how evaluations of one-to-one computing have been undertaken to date and lessons learned about evaluations, although we do also note key insights on one-to-one computing arising out of the now extensive number of evaluations undertaken to date. As such, in appraising or evaluating evaluations, we adopt a meta-evaluative stance (Finn, Stevens, Stufflebeam, & Walberg, 1997; Rodríguez-Campos, 2004; Stufflebeam, 2000a, 2001). A comprehensive meta-evaluation framework for how to assess evaluations of education programs has been developed by 16 professional associations (Sanders & The Joint Committee on Standards for Educational Evaluation, 1994). They identified evaluation principles and four attributes of quality evaluations: utility, feasibility, propriety, and accuracy.

Utility standards address the information needs of those who will use the findings from the evaluation and include standards dealing with stakeholder identification, information scope and collection, evaluation impact, and others. Feasibility standards are concerned with an evaluation being realistic, prudent, diplomatic, and frugal. For example, two feasibility standards are political viability and cost effectiveness. Propriety standards refer to the evaluation's legal and ethical obligations; these include service orientation, rights of human subjects, and conflict of interest, among others. Accuracy standards address the trustworthiness of evaluation data and include, among others, context analysis, valid information, and reliable information.

In appraising evaluations, we address three related questions:

- What do we know about evaluations of one-to-one initiatives based on evaluation studies to date?
- How does what we know about one-to-one computing from evaluations relate to the wider field of research on educational technology?
- What do we need to know in the future in terms of the design and evaluation of one-to-one initiatives?

In addressing these questions we draw upon a range of studies, including evaluations of small- and large-scale one-to-one computing initiatives, research syntheses (Andrews, 2004; Murphy et al., 2002; Penuel, 2006; Murphy, Penuel, Means, Korbak & Whaley, 2001; Vahey & Crawford, 2002), literature on the process of scaling up educational innovations (Coburn, 2003; Dede, Honan & Peters, 2003), and literature on evaluation and research policy (Gardner & Galanouli, 2004; Lagemann, 2000; National Research Council, 2002; Rodriguez-Campos, 2004; Stufflebeam, 2001). While we focus mainly on laptop initiatives because these are the most widely attempted one-to-one computing innovations, we also draw upon evaluations of handheld computing initiatives (e.g., Vahey & Crawford, 2002). Given the significant investment in, proliferation of, and ambitious aims of laptop initiatives and other one-to-one initiatives, many have included an evaluation study (Penuel, 2006). In order to illustrate the complex and multifaceted nature of one-to-one programs, we use the story of TC Williams High School, whose laptop program developed in anything but a linear fashion.

The case of laptops in TC Williams High School: street level and official policy makers all evaluate

In December 2006 the *Washington Post* published an article recounting the evolution of a laptop program over the 2 years since each student at TC Williams High School, Alexandria, Virginia, had been given a laptop. In 2004, district education officials heralded the laptop program as a bold leap into the future and a means of bridging the digital divide. The storyline thereafter was not a simple linear progression or an uncontested adoption of the laptops. Rather, most students left laptops at home unless mandatory for a class or period of the year. One 18-year-old senior student quoted in the article said, "Mine was pretty much under my bed all last year, except for one time a quarter, when it was mandatory. I thought it was just a pain to have to lug it to school."

Furthermore, some teachers were unconvinced about the worth of the program and called the laptops "expensive paperweights." Some school board officials "questioned the academic payoff of a $1.65 million annual expenditure." One school board official whose daughter was a student in the school claimed that "the decision was made to bring computers into the school system before they really knew what they were going to be doing with them." Even the new award-winning school principal had serious questions about the program; he summed it up by describing his main observation on arriving in the school: "The biggest thing I had is, why aren't the kids using them more?" Many other officials also began to ask if the laptops were doing anything to improve students' test scores.

With the goal of making the laptops indispensable, under the leadership of the new principal and with district support, the school set up a range of supports to integrate laptops into teaching and learning, including extension of the school's library hours to facilitate using the Internet for research (the entire campus is a wireless hot spot, with filters to block inappropriate Web sites) and adoption of Blackboard as an online virtual learning environment (VLE). The VLE was meant to ensure that students "use their laptops to participate in class discussions, organize work, take tests and check assignments, among other things." The district also added six full-time positions to support its district-wide program, which, by the end of 2006, involved leasing 3,400 laptops for students and teachers.

There are a number of salutary lessons for understanding one-to-one computing evaluations from this account of the fate of the laptop program in TC Williams High between 2004 and 2006:

• Because computers are a high-profile and visible intervention in schools, many people—certainly more than just official evaluators—are involved in their actual evaluation. Students, teachers, school administrators, and school board members alike all undertook and acted on their street-level evaluation. As Weatherley and Lipsky (1988) remind us, all stakeholders in an innovation—that is, what they term "street-level policy makers"—undertake policy and evaluation.

• From an evaluation perspective, the fate of the laptop program could not have been understood without getting close to the unfolding story in the school. Reliance on administration of standardized pre- and post-tests would have missed the stop–go nature of the program and the various levels of resistance, mixed degrees of belief, and interlinkage of supportive factors that eventually put the program on a firmer footing in the school.

• The extent to which laptops improve student achievement is never far from the agenda (Penuel, 2006). From an evaluation perspective, designing studies that will provide sufficient evidence to support claims about laptops directly enhancing student achievement is a big challenge for a range of reasons, including the fact that laptops (or other one-to-one digital tools) are open to myriad uses and interpretations, the skills of good laptop use may have beneficial impact upon such student engagement in learning, and so-called "twenty-first century skills" (e.g., searching for, filtering, and communicating knowledge) may be hard to measure (Means & Penuel, 2005;

Warschauer, 2006). These latter skills may be more easily assessed using more holistic and authentic assessments than are used in the small number of outcome studies on laptop programs to date.

- The TC Williams High story cautions resisting any temptation to view laptops as a magic bullet, and presses for careful attention in both the design and evaluation of laptop programs to the manner and variety of the initial inputs, processes, and outcomes. That is, what is the best mix of goals, support, and resources with which to get a return on the district's investment and enhance student achievement?

*Evaluation foci: return on investment and
enhancing student achievement*

Evaluation of the one-to-one computing phenomenon is vitally important for a number of reasons, including the initial start-up investment costs, the ongoing expenditure in supporting programs, the ambitious aspirations justifying the investment in terms of both proximal (e.g., improving achievement in students' reading and writing) and distal (e.g., developing digital citizenship for twenty-first century knowledge economy) goals, and the diverse ways in which one-to-one computing has been enacted in various contexts.

The impact of one-to-one computing fits within a wider debate about the return on investment and outcomes on learning and other aspects of schooling due to technological innovation (e.g., Cuban, 2001; Fuchs & Woessmann, 2003; UNESCO, 2003; Warschauer, 2006; Zhao & Conway, 2001). For example, as we have argued elsewhere in this book, one-to-one computing is not a panacea. Rather, we have made a case for a co-evolutionary view of technology in schools (Lei, 2005; Zhao & Frank, 2003). As such, the actual impact of technology on student learning in specific subject/content areas in terms of return on investment is a hotly debated topic and provides some insights on key issues worth addressing in the context of one-to-one computing, such as the nature of studies being undertaken (e.g., data on student achievement across contexts at two or more time points analyzed using multilevel modeling).

Returning to two sets of data that we noted earlier illustrates the contested nature of outcomes. First, large-scale systemic studies of the impact of computing on student attainment (e.g., the Impact 2 study in the United Kingdom [Pittard, 2003]; PISA-based study in 31 countries [Fuchs &

Woessmann, 2003]; a major review of the ICT impact in schools undertaken in the United Kingdom [Condie & Munro, 2007]) point to the moderate or neutral impact of computing on achievement. Other large-scale studies use nationally representative data sets (e.g., U.S. students' performance in the NAEP (National Assessment of Educational Progress) on the 2001 history assessment; see Wenglinsky, 2005, 2005/2006).

Pittard's study points to the moderate impact on student attainment, given supportive conditions. Similarly, Wenglinsky's NAEP-based studies in the United States point to how the optimal role for technology depends on its actual use for students at a particular grade level and for given subject/content area. For example, his 1998 study demonstrated that for younger students, the quality of teacher-directed computer use—defined as a focus on higher order skills rather than drills on routine tasks—impacted student attainment in reading, mathematics, and science. For high school students (2001 study), using computers for generic academic tasks (e.g., word processing, using computers for projects, creating charts and tables, and communication via e-mail and chat rooms) predicted higher student achievement than use for subject-specific tasks such as reading primary source documents.

Wenglinsky's studies, in particular, point to what we might term "goodness of fit" between computing and curriculum in the classroom. That is, generic claims about the impact of computing need to be moderated by an understanding of how computers fit within specific curriculum and educational niches. Thus, while it may be of interest to know the average effect of a one-to-one initiative, other contextual information is needed by policy makers and practitioners seeking to design and plan under local conditions. What types of evaluations have been undertaken in the case of one-to-one computing and how useful might they be in the future? In order to address this question, we turn to a number of key laptop program evaluations.

What do we know about the evaluation of one-to-one computing?

In relation to one-to-one computing, there have been many different evaluation contexts, including large-scale district level evaluations (Zucker & McGhee, 2005), small-scale district level evaluations (Johnstone, 2003b; Stevenson, 1999), state level evaluations (e.g., Harris & Smith, 2004; Silvernail & Lane, 2004; Urban & Zhao, 2003), evaluations of small-scale national pilot projects (Conway, 2005), and evaluation of large-scale nationwide initiatives (Rockman et al., 1997, 1998, 1999, 2000).

A common feature of laptop evaluations to date is that they have typically been global descriptive studies of implementation rather than quasi-experimental or experimental outcome designs (Penuel, 2006). A small number have incorporated pre- and post-test measures of student achievement in one or more core curricular areas (e.g., Gulek & Demirtas, 2005; Harris & Smith, 2004; Johnstone, 2003; Rockman et al., 2000). The 2005–2006 evaluation of Michigan's Freedom to Learn (FTL) laptop program involves a global descriptive component as well as a quasi-experimental design comparing 10 FTL schools and 10 matched control schools. One of the notable strengths of a number of these evaluations is that they have been longitudinal and tracked developments of laptop initiatives over 2 or more years (Conway, 2005; Gulek & Demirtas, 2005; Rockman et al., 2000; Warschauer, 2006; Zucher & McGhee, 2005).

For the purposes of this chapter, we adopt the evaluation model used by Zucker (2004) and Zucker and McGhee (2005) in reporting on the Henrico County School District laptop program; with its emphasis on context and inputs (i.e., critical features) as well as processes and products (i.e., interactions as well as intermediate and ultimate outcomes) it is a typical evaluation model (Stufflebeam, 2000b) (see Figure 7.1). Zucker and McGhee's research and evaluation model is valuable for a number of reasons. In particular, it highlights the potential for distinct differences in one-to-one computing initiatives in drawing attention of evaluators to the critical features of one-to-one initiatives (e.g., hand-held, mobile phone, or laptop). The emphasis on intermediate and ultimate outcomes is also valuable, with the former drawing attention to some of the conditions we have discussed in earlier chapters.

Aims of one-to-one laptop programs in policy discourse

Recognizing the hope invested in laptops and other one-to-one initiatives as a vehicle to foster ICT and technology integration, improved motivation to learn, greater equity, increased academic attainment, and even so-called twenty-first century skills, it is no surprise that, as Schaumburg notes, "the use of mobile computers has spread worldwide" (2001, p. 1), with numerous laptop and other one-to-one initiatives undertaken over the last decade in various education systems

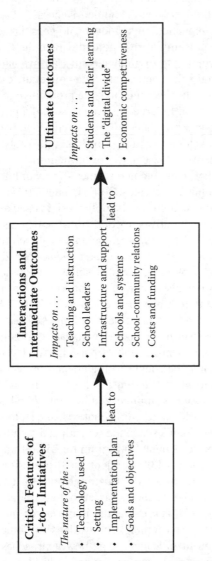

A Framework for Research and Evaluation of One-to-One Computing

Critical Features of 1-to-1 Initiatives

The nature of the . . .

- Technology used
- Setting
- Implementation plan
- Goals and objectives

lead to

Interactions and Intermediate Outcomes

Impacts on . . .

- Teaching and instruction
- School leaders
- Infrastructure and support
- Schools and systems
- School-community relations
- Costs and funding

lead to

Ultimate Outcomes

Impacts on . . .

- Students and their learning
- The "digital divide"
- Economic competitiveness

Figure 7.1   Zucker and McGhee's one-to-one computing evaluation model.

worldwide. The plethora of aims associated with laptop initiatives has complicated the evaluation process. For example, how will an evaluation provide evidence that a 1- or 2-year laptop program will enhance a region's or nation's economic competitiveness?

Even if a reduction in achievement gaps is acceptable as evidence of a laptop initiative's potential to reduce achievement gaps and thereby contribute to greater equity, an evaluation design necessary to undertake the appropriate data collection and allied statistical analysis across school settings may be prohibitive in terms of cost and investment in labor. In large part due to cost and pragmatic evaluation design decisions, while there has been a considerable number of evaluations undertaken on the impact of laptop projects and other mobile learning initiatives, these have overwhelmingly focused on implementation rather than outcome studies (for a summary, see Naismith, Lonsdale,Vavoula, & Sharples, 2005; Penuel, 2005).

First, we want to focus on the policy origins and aims of laptops and other one-to-one initiatives in educational policy discourse. In chapter 2 we identified the various arguments being used to make a case for one-to-one computing. In this chapter, we address these arguments in more depth by looking at aims of some one-to-one computing projects. We start our discussion of evaluation with a brief review of the goals set by different one-to-one laptop projects. The initial impetus for any innovation often has embedded in it key assumptions and aims that will more than likely shape important decisions about choice of technologies, the identification of participants, the type of evaluation evidence that may be sought to understand project development, and the audiences to whom "findings" will be addressed (see Table 7.1).

The various studies summarized in Table 7.1 illustrate the range of policy origins, different teaching contexts, and laptop initiatives of very different scales, from a statewide seventh and eighth grade program in Maine and Henrico County's program for 25,000 students to a national pilot project involving 1,000 students in 31 postprimary schools in Ireland. As we noted earlier, the Birmingham ALL one-to-one initiative in the United Kingdom encompassed both proximal and distal goals, and this dual emphasis is evident in many laptop initiatives.

Furthermore, for example, innovations in some curricular areas rarely excite business leaders (e.g., new approaches in music or art); however, because it is seen as a potentially high-yield innovation in developing knowledge economies, technological innovation excites the interest of business and government. Such enthusiasm from the business world may result in economy-related goals being linked to one-to-one

Table 7.1   Selected one-to-one initiatives: policy origins and teaching and learning contexts

Initiative	Policy origins	Teaching and learning contexts
Maine laptop program	Assist Maine to leapfrog economically into twenty-first century; Act as a test bed for a state level laptop program	All grade 7 and 8 students in state
Henrico County School District, Virginia (Zucker & McGhee, 2005)	Act as a test bed for a state level laptop program	25,000 grade 6–12 teachers and students
Laptops initiative for students with dyslexia and other reading and writing difficulties, Ireland (Conway, 2005; Daly, 2006; SENJIT, 2006)	Address literacy needs of students with literacy learning difficulties	1,000 grade 8–10 postprimary students in 31 schools around the country
Freedom to Learn (FTL), Michigan (Urban & Zhao, 2003)	Bridge rural–urban digital divideImprove student achievement in core academic subjects	20,000 middle school students and 1,200 teachers from 188 schools in 95 districts
Maine and California case studies (Warschauer, 2006)	Reform schools; Bridge achievement gaps; Increase test scores	Case studies in 10 schools
Birmingham (England) Anywhere, Anytime Learning, BGfL	Raise academic standards; Take advantage of shifting employment patterns; Develop twenty-first century skills	Selected schools in city with plans to extend to all students

computing innovations, even if connections between increasing uptake of one-to-one computing and economic output are inspirational, without any clearly identifiable or testable relationships.

The largest and most well documented laptop project, Microsoft's "Anytime, Anywhere" learning initiative, also had ambitious goals focused on how the project would transform students' educational experiences (Rockman et al., 1997, 1998, 1999). We note some of the key policy assumptions underpinning this initiative as it has not only been a notable project but its title, "anytime, anywhere" has also been the conceptual core of many other laptop policies. Thus, Microsoft's project represents a powerful symbolic force in the drive to integrate laptops in schools. The promise of omnipresent and ubiquitous computing underlying the Anywhere, Anytime project sent significant ripples through the educational world, but especially through ICT and educational technology circles at the policy level. For example, "anywhere, anytime" is the title of a Massachusetts adult basic education program (see http://anywhereanytimeabe.org/). Even if the creators of this adult education technology initiative did not directly draw on Microsoft's program for schools, the use of "anywhere, anytime" speaks to the inherent appeal and rhetorical sophistication of those who first coined the term by joining notions of omnipresence and ubiquity to a memorable phrase that continues to reverberate around the educational world.

For example, San Lorenzo School District in California describes an Anywhere, Anytime Learning partnership as "a grant program for asynchronous, innovative, scalable, and nationally significant distance education projects. Eligibility requirements for LAAP include at least two partners and a one-to-one ..." Walled Lake Elementary School in Detroit, Michigan, documents an Anywhere, Anytime Learning (AAL) project "that combines leading-edge wireless technology with laptop computers to allow students in grades 5–10 the chance to have a personal laptop computer that they can use anytime," noting that there "are over 2,000 students currently participating in the program." Each of these projects, whether directly or indirectly inspired by the Microsoft's Anywhere, Anytime Learning laptop initiative, testifies to the appeal of the "anywhere, anytime" concept. Paying attention to laptop program policy origins' aims is important in appraising the appropriateness of evaluation approaches, evidence presented, and conclusions drawn across different studies, and their suitability for informing other contexts.

Key evaluation findings from implementation and outcome studies

In this section, using Zucker and McGhee's (2005) three-part evaluation framework to organize findings, we summarize key findings from evaluation studies on one-to-one computing. In doing so, we draw in particular upon a key research synthesis undertaken by Penuel (2005, 2006), major studies of handheld computing (SRI, 2002), and other significant studies of one-to-one initiatives (e.g., Henrico County Public Schools Project [HCPS; Zucker & McGhee, 2005; Davis, Garas, Hopstock, Kellum, & Stephenson, 2005]; Michigan FTL Project [Urban &

Zhao, 2004, 2005]; Laptops Initiative, Ireland [Conway, 2005; Daly, 2006; SENJIT, 2006]; Chilean Pocket PCs With Wireless Network Program [Nussbaum & Zurita, 2005; Warschauer, 2006a, 2006b]; Alpha Middle School 1-1 Laptop Project [Lei, 2005; Lei & Zhao, 2005]; see Appendix 1 in this book).

Implementation and effects of one-to-one computing initiatives—research synthesis (Penuel, 2005, 2006)

Penuel's research synthesis, in our view, is a particularly valuable review of the laptop phenomenon in education as it was very comprehensive in identifying 245 research studies in the initial search and, after selection based on research rigor and scope of studies, involved a synthesis based on 46 of these 245 studies. Apple Computer's funded research synthesis, which only focused on evaluations of wireless one-to-one laptop initiatives undertaken between 2001 and 2005, has many of the hallmarks of a meta-evaluation (Stufflebeam, 2001) in that it deals with some aspects of utility (e.g., information needs of those who will use findings), feasibility (e.g., degree to which one-to-one initiatives are realistic and evaluations match stated project goals), and accuracy (e.g., political dimension of investment and the cost effectiveness of one-to-one computing initiatives).

There has been a considerable increase in the number and quality of evaluations of one-to-one computing over the period 2001–2005 (Penuel, 2005). In an earlier review for SRI (Penuel et al., 2001) the researchers then noted the dearth of studies at that time and the relatively weak evaluation designs being used. Furthermore, Penuel (2006) claims that, by 2005, an increasing number of one-to-one computing evaluation and research studies were being published in refereed journals compared to 2001, when none were available (i.e., Jaillet, 2004; Newhouse, 2001; Newhouse & Rennie, 2001; Trimmell & Bachmann, 2004; Warschauer, Grant, Real, & Rousseau, 2004; Windschitl & Sahl, 2002).

Implementation studies (Penuel, 2005, identified 27) rather than outcome studies are the predominant type of evaluation undertaken. Both outcome and implementation studies that adopted systematic data collection and analysis procedures were included. Implementation studies use survey, interview, and case study data to describe projects in order to promote understanding of project implementation processes in terms of the perceptions of various

Table 7.2   Implementation and outcome studies

	Implementation	Outcome
Goal	To understand project evolution from perspective of participants	To identify the relative influence of variables contributing to project outcomes
Design and data collection	Observations, interview, and surveys	Experimental design with intervention and treatment groups

project participants such as teachers, students, and parents, as well as project funding agencies and project managers. Outcome studies use experimental designs with random assignment or quasi-experimental designs with pre- and post-test data for both intervention and comparison groups.

Drawing upon Penuel's (2006) research synthesis and other major evaluations noted before, we highlight key findings in relation to the design of one-to-one computing research and evaluation studies.

Types of studies

Many implementation studies and few outcome studies. Of the 46 studies used in the final research synthesis in Penuel's review, only 3 were outcome studies (i.e., Gulek & Demirtas, 2005; Russell, Bebell, & Higgins, 2004; Schaumburg, 2001). As such, outcome studies are the exception rather than the rule because they have hardly been used at all to date. Outcome studies adopt objective measures of project impact, typically involving intervention-comparison group designs (see Table 7.2). As such, almost all one-to-one evaluations were implementation studies focused on describing impact based on teacher and/or student report (often over an extended period of time), but they did not collect pre- and post-test student achievement data. The 2005–2006 evaluation of Michigan's FTL laptop program involves both a global descriptive component as well as a quasi-experimental design comparing 10 FTL schools and 10 matched control schools. One of the notable strengths of a number of the implementation evaluations is that they have been longitudinal and tracked developments of laptop initiatives over 2 or more years (Conway, 2005; Daly, 2006; Lei, 2005; Rockman et al., 2000; Urban & Zhao, 2004, 2005; Warschauer, 2006; Zucker & McGhee, 2005).

Critical design features of one-to-one initiatives

- *Many proximal and distal goals.* The goals of one-to-one initiatives vary considerably but typically encompass both proximal and distal aims (Warschauer, 2006). According to Penuel (2006), many focus on one or more of four desirable outcomes: enhancing academic achievement, promoting equity, increasing economic competitiveness, and transforming classroom teaching.
- *One-to-one initiatives included diverse technologies, but laptop projects were most frequent.* While there have been a considerable number of handheld projects (see Vahey & Crawford, 2002, for a review of 102 projects in the United States), laptop projects are the most widespread and significant one-to-one innovation.

Interactions and intermediate outcomes

- *Teachers go through stages of technology integration.* Most studies indicate that teachers are in the early stages of technology integration, or what Sandholtz, Ringstaff, and Dwyer (1997) term "adoption." Many studies have documented that teachers appear to go through stages integrating technology. In the case of one-to-one computing, this seems to translate, according to Penuel (2006), into teachers' focus on basic skills and use of productivity in the adaptation stage. As teachers become more familiar with and flexible in their use of laptops, they typically use them for more advanced skills such as projects and problem-solving tasks (Penuel).
- *Five excellent reasons to start a laptops project.* Warschauer (2005/2006, p. 35) identified, what he called "five excellent reasons to start a laptop project" based on a 2-year case study in 10 Maine and California schools: twenty-first century learning skills, greater engagement through multimedia, more and better writing, deeper learning, and easier integration of technology. Warschauer's focus on benefits of laptop programs is echoed by Penuel, who noted that, when more rigorous designs were used, the available evidence base, "is

generally positive, especially with respect to laptop programs' effects on technology use, technology proficiency, and writing skills" (2005, p. 13).

- *Teachers matter.* Teachers matter on a number of levels, including their beliefs about technology, technical competence, and the extent to which they have a positive attitude about technology (Penuel, 2006). Teachers' beliefs and attitudes have a significant impact on implementation, as McGrail (2006) exemplified vividly in a study of a laptop initiative in a secondary English setting. McGrail's study documents the many ways in which teachers' perspectives on one-to-one technologies are reflected in their use, nonuse, and type of use in classrooms. According to McGrail:

Unfortunately, however, when asked to describe their overall experiences and attitudes toward technology, these teachers revealed a great deal of ambivalence about it in their instruction, especially in the context of a school-wide laptop technology initiative. Four larger clusters of conflict contributed to this ambivalence: (1) conflicts surrounding institutional control in implementing the laptop program and teacher agency; (2) conflicts surrounding standardized testing's uncertain relationship with technology mandates; (3) conflicts surrounding technology uses in the general curriculum and technology allocation in specific class types; and (4) conflicts surrounding professional identity and the challenges that both student and teacher technology use brought to this identity. The study concludes that these teachers needed to be given greater agency in planning and implementing the laptop technology initiative and in revising their curriculum to embrace this new technology, and the necessary professional development to prepare them for such an educational innovation. (p. 1055)

- *Professional development matters.* Professional development and technical support are critical for implementation. On-going professional development of teachers emerges as an issue of central importance across many studies, and Penuel (2006) highlights this as one of the key findings of implementation studies.

Ultimate Outcomes

- *Evaluation and study design matters.* Studies with more rigorous designs show more positive effects. Penuel (2006) cites the case of a one-to-one project in a single elementary school that involved three types of laptop deployment: four to one, two to one, and one to one. The evaluation (Russell et al., 2004) indicated that the 1:1 model had many advantages over the other two models. That is, the 1:1 classrooms were characterized by greater use of computers across the curriculum and greater use at home for academic purposes; more often than not their writing included the use of computers. Furthermore, Penuel claimed that the clearest evidence of impact emerged from the best designed studies.
- *Lack of outcome studies.* There is little implementation of well-designed outcome research studies on one-to-one projects. Penuel (2006) only identified three that met such criteria in relation to wireless one-to-one computing. The lack of outcome studies is problematic, given one-to-one computing, because these are essential to evaluate the extent to which it enhances student achievement. As such, if improving student achievement is espoused as one of the primary aims of laptop initiatives, it is surprising that evaluation designs do not address this matter. However, as Gardner and Galanouli note, "…it is salutary to remember that much of the work to date, including that of the few large-scale projects such as ImpaCT2 (Harrison, Somekh, Lewin, & Mavers, 2003), has been relatively inconclusive about the nature of the impact that ICT has had" (p. 151).
- *Strengthening measurement of hoped-for outcomes.* Strengthening measurement of hoped-for outcomes will lead to improvement in future research on one-to-one projects. As such, it is critical that the particular aims of one-to-one programs are reflected in evaluation designs that can appraise, for example, whether and how they have enhanced students' digital literacy skills and other twenty-first century skills (e.g., higher order thinking, problem solving, searching for, organizing, and communicating information).

Overall, it seems reasonable to conclude that evaluations of one-to-one initiatives are not as strong as might be hoped, especially if evidence-based practice is seen increasingly as an important feature of the educational policy landscape. Some doubt its ready applicability in education, given its more uncertain and indeterminate context than, for example, medicine, where cause and effect may be more easily determined, at least in relation to physiological matters (Gardner & Galanouli, 2003). Lamenting the dearth of strong evaluation studies, Penuel (2005) concluded that, in relation to the quality of existing research designs, "The fact that here are few studies of laptop programs that measured both outcomes and implementation means that there is a weak research-based evidence base for deciding what the critical components of laptops programs really are" (p. 9).

Penuel's research synthesis thus makes sobering reading for those who might seek to advocate a one-to-one program drawing upon existing research as evidence of how to implement a project or what outcomes might be reliably expected. As such, Penuel's study is immensely valuable in providing a measured and comprehensive review of the quality of the research-based evidence available on one-to-one computing. However, while we agree with the need for rigorous designs, we would put less emphasis than Penuel (2006) on quasi-experimental and experimental study designs as the predominant means of enhancing rigor. In response to what we see as an overemphasis on quasi-experimental designs, we make a case for an ecological approach to the study of technology integration in schools. This approach assumes a reciprocally determining rather than causal explanatory model and may fit more powerfully with the manner in which various factors interact in one-to-one projects. As such, we want to emphasize the importance of the research discourse in shaping future one-to-one evaluations.

One-to-one computing evaluations and educational technology research

Beyond the "technology as magic bullet fallacy": getting the "proper mix"

In discussing the evaluation in the context of one-to-one computing and the development of ubiquitous learning environments, we situate this within earlier discussions about the interwoven nature of school context and conditions for developing ubiquitous learning environments (chapter 3). At this point, we also want to note the "technology as magic bullet" fallacy, which has plagued research on educational technology in successive waves of technological innovation over the last 100 years (Clark, 1983; Cuban, 2001; Michaels, 1990; Selwyn, 2003). Nevertheless, there is increasing attention to contextually sensitive models of technological innovation in schools (e.g., Venesky & Cassandra, 2002; Warschauer, 2006; Zhao et al., 2003).

In order to emphasize the manner in which technology cannot be a magic bullet but rather may be part of a changing school ecology, we draw on research undertaken in the 1980s by researchers alert to and wary of the linear logic attributing all-powerful change to the mere introduction of a particular technology. Research on the influence of the media on learning has been a consistent feature of educational research for almost 100 years. Thorndike (1912), for example, recommended pictures as a labor-saving device in classroom teaching (Clark, 1983). As Clark noted, "Most of this research is buttressed by a hope that learning will be enhanced with the proper mix of medium, student, subject matter content, and learning task" (p. 445).

However, the so-called proper mix cannot be understood without taking context into account, as Clark notes. For example, citing the introduction of TV in El Salvador, he claims that it was "not the medium that caused change [but] rather the curriculum reform that accompanied the change" (p. 445). Roschelle, Pea, Hoadley, Gordin, and Means (2000) make a similar point in their review of effective educational technology initiatives in the United States. However, this attention to the wider context of technological innovation has been on the minds of some researchers in the field, even if this perspective has not been heeded at times.

For example, in an insightful study on the issue of how context redefines educational technology innovations, Michaels (1990) questioned the causal logic underpinning the introduction of computers into schools by examining the use of microcomputers for writing in sixth grade classrooms. The initial research question was, "What impact will computers used for writing have on life in classroom, teacher-student interaction, and student literacy?"

Education researchers, teachers, and software developers alike are aware of the pressing need to assess the impact of microcomputer technology on student learning. However, in discussions about this new technology, it is generally assumed that a given computer with a particular kind of software will have a specifiable and generalized impact on classrooms, teachers, and students. That is, the computer tends to be thought of and studied as an independent variable, as a controllable and quantifiable agent of change. (Michaels, 1990, p. 246)

Thus, Michaels convincingly argues that an unstated and unexamined assumption was that, by introducing the same computer—the Apple—to each classroom, the writing software program QUAYLE, the same technology would produce similar change into both classrooms. However, this assumption proved to be incorrect. Assumptions that the new technology would be key in reshaping the learning environment in the two rooms was overturned in favor of a hypothesis that the computers themselves were shaped to fit the already established patterns of social organization:

> Because the two learning environments differed, the same computers with the same writing software ended up being used differently, and came to serve as different writing tools. For this reason, we have come to think of the computer as the dependent variable, and [it] is itself affected by the classroom context, and then in turn, having an influence on it. (Michaels, 1990, p. 246)

The author argues that a one-way causation model—either the computer causes a change in the social setting or a social setting causes a change in computer use—is unsatisfactory, and she opts for a theory of computers and social settings that views the relationship as mutually constitutive (i.e., co-evolution). However, the author titled the article "The Computer as Dependent Variable" (i.e., dependent variable = outcome) in order to bring into question the simplified model and a set of frequently unexamined causal assumptions underlying much of the research on the impact of computers in classrooms. Over the course of the ethnographic study, Michaels developed an analytic construct to understand the relationship between computers and students' written products. She called this construct the "writing system," by which she meant:

> the activities, norms, the rights and obligations for speaking and acting, and uses of technology that influence and constrain students' writing in the classroom. As we use the term, the writing system is the day-to-day practice of a curriculum, shaped largely by the teacher, partly by the students and partly by outside forces that impinge on the classroom. (p. 247)

Perhaps one of the most interesting observations Michaels makes on the outcomes of this 3-year study is that computer entry, or technological innovation, came to be seen as a small component embedded within a larger social system rather than as the major and most important factor of educational change. Furthermore, Michaels notes that if we want to understand the impact of computers on classrooms and curriculum more fully, we must see the computer "as influenced by and influencing the past and context; as a dependent variable and independent variable" (p. 254). In this light, commenting on the role of teachers, Michaels concludes that "most importantly, teachers need support in becoming critics and evaluators of their own pedagogical goals, of patterns and practices in their own settings, and of the potential of technology in light of their goals and strengths as teachers" (p. 254).

In light of the insights of Clark (1983) and Michaels (1990), we note how the views of both are largely consistent with the emphasis we have placed on context as we argue for an ecological perspective on one-to-one computing and the development of ubiquitous learning environments. That is, we agree with Michaels that one-to-one computing is not an independent variable driving change, and we agree with Clark that getting the proper mix of learner, teacher, content, teacher learning, technology support, and one-to-one computing must take place with an eye on both the micro- and macrocontext of such innovations. Attending to the proper mix in context demands particular approaches to evaluation and it is to this we now turn. Clark's and Michaels's perspectives also draw attention to the central importance of educational change models in analyzing technological innovation.

Evaluation and research discourse matters

The high value and significant investment in randomized controlled trials (RCTs) approaches to educational research over the last few years has shaped the context within which evaluations of educational technology innovations take place (Lagemann, 2000; National Research Council, 2002). This policy shift toward a positivist epistemology and research designs congruent with its assumptions have pressed evaluators and others involved in educational reform efforts to engage in more careful attention to the nature of evaluation designs and strengths and limitations of chosen approaches.

However, few, if any, evaluations can live up to the rigorous design standards demanded within the RCT model of research. In relation to evaluations of one-to-one initiatives, the dominance of implementation rather than outcome studies is far short of the rigor demanded from an RCT perspective (Penuel, 2006). Implementation studies, from the perspective of those advocating RCTs as the "gold standard" in educational research design would argue that one-to-one initiatives are being advocated for based upon a poor evidence base. More broadly in the context of educational technology research, the relatively small number of studies that meet criteria for inclusion in meta-analyses in major reviews of e-learning in the United States (Murphy et al., 2002) and the United Kingdom (Andrews et al., 2007) attest to the difficulties in advancing knowledge in the field—given the dearth of studies that meet criteria specified in RCTs and related research design principles being adhered to by research review teams in organizations as such as SRI in the United States and Evidence for Policy and Practice Information (EPPI) in the United Kingdom.

For the purposes of our work, we note some of the findings from the relevant reviews of studies (e.g., Murphy et al., 2002; Andrews et al., 2006) and findings from evaluations. While meta-analyses and reviews of one-to-one computing studies tell us a lot about how initiatives impacted learning and other outcomes in various ways, they often have less to say about why these studies worked and how one-to-one programs support more distal goals. It is to this we now turn. We note two reasons offered by Cuban to explain the sluggish integration of technology in schools and offer an ecological model as a third option.

Theories of educational change and technological innovation: models of change matter

In his now prominently cited research (e.g., Haddad & Draxler, 2002, p. 145; UNESCO, 2003, p. 26) on the scope and nature of technology integration in schools, Cuban (2001) used Sandholtz and colleagues' (1997) five-stage model of technology integration to assess the manner in which computers were integrated into the daily fabric of teaching and learning in early elementary school, high school, and university classrooms in Silicon Valley, California, during the late 1990s. In an effort to explain the poor uptake of technology he offers two possible explanations: "slow revolution" and "history and contexts" theories of educational change.

Before we address these, we provide the context for his theories of change because they are centrally important in offering an alternative: an ecological model. In the case of early childhood, Cuban studied seven school sites. Seven of the eleven teachers at the seven sites were at the adoption stage, three were at the adaptation stage, and one was at the appropriation stage. In summarizing the findings of his study, Cuban commented as follows:

> To fervent advocates of using technology in schools, no revolution has occurred in how the teachers organize or teach in these classrooms. Nor have there been dramatic or substantial changes in how teachers teach or in how the children learn. If anything, the addition of a computer center ... means that teachers have adapted the innovation to existing ways of teaching and learning that have dominated education for decades ... if anything, teachers' limited use of computers signalled ambivalence, even their uncertainty over the proper use of technology for children. (2001, p. 59)

Introducing his intensive study of two California high schools, Cuban criticized findings and claims about the nature of computer use in schools as follows, "What I find in the national data is far too much reliance on self-reports and far less investigation of actual use in local schools" (2001, p. 73). Drawing upon both observation and interview data with 35 teachers and 33 students in these two typical schools, he concluded:

> Based on what we saw and what teachers and students reported, we concluded that the integration of computers into a classroom curricula and instruction techniques was minimal. It ranged from entry-level to adoption, with fewer than five of the adaptation level. We note this is only for academic teachers in both schools—excluding those teachers designated to teach computer classes—who effortlessly and continually use technology in their classes, appropriate as it is as part of their weekly work. (2001, p. 90)

How can we interpret Cuban's study in relation to one-to-one programs? First, as Cuban noted in the introduction to his book, if there is anywhere in the world one might expect computers to be widely and regularly used in classrooms it is in Silicon Valley, California, since no other setting is so culturally and materially well disposed to the favorable integration of computers in schools. As such, Silicon Valley can be seen as a school context in which there is abundant access to new educational technologies, and it thus provides, he claims, a powerful and compelling test case of technology integration.

Second, the fact that his study examined technology integration at elementary, high school, and college levels of the education system helps us to understand the structural and cultural features often shared across levels of educational establishments that may inhibit, or at the very least, significantly slow down the pace of technology integration. Cuban's position on the actual integration of technologies into classrooms and the future likelihood of technology integration can be conveyed by attention to a number of cleverly titled sections and chapters in his book: "High-Tech Schools, Low-Tech Learning"; "Maximal Access, Minimal Change"; and "New Technologies in Old Universities."

Third, in a wide-ranging and provocative fifth chapter discussing these somewhat expected outcomes, Cuban evaluates the potency of two different theories that might explain the unexpectedly low levels of technology integration found across three levels of California's education system. The first he identifies as the *slow revolution* explanation, and the second has the *history and contexts* explanation. This slow revolution argument explains the lower levels of technology integration in terms of evolution rather than the revolution. Critical of the overblown revolutionary rhetoric of some technology advocates, slow revolution advocates see technology integration occurring incrementally over years and decades rather than over weeks and months or even a full school year. From this slow revolution perspective, Cuban notes that it has taken 100 years for plane flight to transform the way in which human beings travel around the world.

The history and contexts explanation "emphasizes the societal role that schools perform in a democracy, the structures and work educators perform, and the symbolic and actual nature of the technological innovation" (2001, p. 156). In relation to the perceived role of schools in society, Cuban notes that it is political suicide for politicians and school leaders not to want to adopt one of the most powerful symbols of modernity—that is, "high-tech" technological infrastructure. Thus, from this perspective, "even with this reluctance that investments in information technologies raise test scores or promote better teaching, most school managers use the rhetoric of technological progress to establish legitimacy with their patrons and the private sector" (2001, p. 159).

Even though Cuban's research has been meticulously undertaken, contextualized at the school and community levels, and reported in rich and detailed case studies, it had a cross-sectional rather than longitudinal focus. Sandholtz and colleagues' (1997) ACOT study, however, had a longitudinal focus and this may be particularly important in drawing attention to how technology integration may change over time in particular school settings. Preliminary evidence from the evaluation of the laptop initiatives and other one-to-one projects suggests that schools evolve considerably in their use of the one-to-one computing even in a relatively short period of time (Conway, 2005; Rockman et al., 2000; Urban & Zhao, 2004; Zucker & McGhee, 2005). What is not clear from Cuban's two theories is how any change can occur at all in the context of one-to-one computing initiatives. In a sense, Cuban offers a theory of educational stability rather than educational change.

However, we also note that, while Cuban is most often cited to support claims that technology does not work in ways it was intended, he did not rule out the possibility that it might be different in the future. This latter point is rarely made in references to his book, *Oversold and Underused: Computers in the Classroom*. For example, "higher teacher and student productivity and a transformation in teaching and learning ... must be tagged as failures. Computers have been oversold and underused, at least for now" (p. 179).

From "what works" to "what works, when, and how":
ecological models of one-to-one computing evaluation

The dominance of RCT discourse in educational research and evaluation has led to an overemphasis on the *what works* question to the detriment of addressing other important and essential questions such as *when* and *how* one-to-one programs work (Means & Penuel, 2005). Due to increasing recognition that policy makers and practitioners need more contextually relevant information, evaluation approaches need to address the "how" and "when" questions. Thus, as recognition of the complex ecology within which one-to-one computing initiatives take place and a greater realization that more rigorous designs are needed to assess intermediate and ultimate outcomes, researchers are beginning to set guidelines to reorient research on technology in schools. This is evident in work by Haertel and Means (2004), who identified principles to guide research on educational technology, and work on scaling up, which attends to the wider educational environment in attempting to spread technological innovation (Dede, Honan, & Peters, 2003).

For example, Haertel and Means (2003) outline key principles that ought to guide evaluation of educational technology in schools as follows:

* Focus on critical influences at different levels of the system.
* Determine and measure key project priorities.
* Seek ways to measure long-term project aims and benefits.
* Ensure the research question drives the choice of method.

Haertel and Means's principles are important in that they highlight the key goals, appropriateness of measures, congruence between research methods and research questions, and the inherent systemic nature of educational technology innovation in schools. For example, the adoption of one-to-one projects is increasingly being linked to the possibility of preparing students for so-called twenty-first century skills. What these entail and how they can be measured in the short and long term are important to address one-to-one projects, as to date such longer term aims have been less central to project evaluations, given the focus on short-term implementation evaluation designs. In advocating for a deeper and more comprehensive approach to educational technology research, Means and Haertel's guidelines are consistent with what we are calling an ecological approach.

First, an ecological model also moves beyond an additive perspective and therefore is increasingly seen as valuable in understanding change and utilizing lessons learned for future planning (Hoban, 2002). As such, additive models tend to focus on a laundry list of factors, whereas ecological models reorient evaluators and those implementing programs to the activities, processes, and practices in classroom, school, and system-wide contexts (Hoban, 2002; Zhao & Frank, 2003). As such, it moves beyond correlational observations (e.g., teachers with this set of beliefs tend to adopt or not adopt certain technologies) and forces attention on activities and practices

Second, in focusing on key processes—for example, the role of teacher networks in supporting or inhibiting change—it highlights the flow of interaction and the reciprocally determining nature of change. For example, even if one teacher relies on another for considerable help as a resource in getting started in using one-to-one computing, the person being helped is also a resource for the person helping in that his or her tacit support, by conforming though participation, is a resource for others in the school.

Third, by focusing on processes and interactions, ecological models can guide the nature of data collected (e.g., social network data; research questions sensitive to interactions, patterns of influence, and the diffusion of ideas). In adopting such methods, while ecological models do not simplify understanding of change, they may provide capacity for greater cross-evaluation review and generation of more powerful models of change as evaluators identify paths of influence and reciprocal determinism in one-to-one computing. For example, one widely supported finding in many one-to-one evaluations is the manner in which additional or ancillary software becomes necessary for deeper and more contextually appropriate implementation of initiatives. Software in this scenario operate as a way for an initiative to adapt to a local ecological niche, and tracking their use may provide an especially valuable understanding of local aims, meaning making, and diffusion through patterns of social influence. As Zhao and Frank (2003) noted in their longitudinal study using a social network perspective of technological innovation:

> Findings of this study suggest that the ecological perspective can be a powerful analytical framework for understanding technology uses in schools. This perspective points out new directions for research and has significant policy and practical implications for implementing innovations to schools. (p. 807)

Such an ecological approach seems to match well with the factors identified as critical in scaling up educational technology innovations (Dede, Honan & Peters, 2003). The researchers, policy makers, and practitioners involved in Dede and Honan's work identified the following set of themes that emerged from their deliberations:

- Innovations must adapt to changing contexts.
- Innovation leaders must adapt to changing funding and policy contexts.
- Innovations require a broad support base.
- Capacity building and partnership ought to be central to project implementation.
- Access to ongoing data in order to inform is essential for project leadership.

We think these themes highlight contextual sensitivity, the need for ongoing review and reframing (rather than linear implementation), the role of local relationships and alliances around some shared understanding of goals, and the role of knowledge in shaping the ecology of one-to-one computing environments. As scaling up one-to-one computing on a system-wide scale has already become a policy issue if not a priority in many parts of the world, we now turn to this question in terms of what is next in one-to-one computing. Furthermore, since these themes from Dede and Honan are valuable and have implications for the future of one-to-one computing policies, we address these also in the final chapter of this book.

The ecological perspective is clearly not neutral in terms of evaluation design. That is, evaluation designs that are contextually sensitive scan for unexpected consequences, address fitness for purpose, attend to issues of scale, and attempt to assess the reciprocally determining nature of social change. However, an ecological perspective does not mean only that evaluation designs have to become implementation focused using only interpretive methods and shy away from results or outcome studies using experimental designs; rather, ecologically oriented evaluation may adopt a variety of methods.

For example, one of the weaknesses of experimental designs is that the "intervention" versus "comparison" group is seen as a dichotomous variable. In educational practice it may make more sense to understand degrees of implementation and develop statistical models appropriate to such a theory of evaluation. In their review of changing evaluation methods, Lipsey and Cordray (2000) identified two statistical methods that we think might add considerable contextual sensitivity to current evaluation that use statistical measures of program impact: (1) analysis of relative group change, and (2) use of hierarchical linear modeling (HLM) to model individual and group change.

Analysis of group change initially focuses on assessing the degree of engagement in the intervention experienced by those intended as the recipients of an intervention as a preliminary step before ascertaining the impact differential between groups. HLM presents a wide menu of statistical models to test a variety of different intervention models (Raudenbush & Bryk, 2002; Sloane, 2005). We might also add a third statistical approach—that is, a social networks strategy (Frank, Zhao, & Borman, 2004). Of course, ecological approaches also seek to tell the story of innovations but do not preclude the use of statistical methods congruent with ecological understandings of individual growth and organizational change.

What's next: framing evaluations of one-to-one computing

When I took office, only high-energy physicists had ever heard of the World Wide Web ... Now even my cat has its own page.

President Bill Clinton, announcing Next Generation
Internet initiative, 1996

Expansion of, what is formally known as, the Maine Learning Technology Initiative is not an option, it is an imperative.

Maine Coalition for Excellence in Education

Fitness for purpose in evaluation studies

Collectively, evaluation studies have contributed a lot to our understanding of one-to-one computing. No individual study has identified a magic bullet, nor is one likely to in the future. However, there are some clear messages emerging in terms of the best mix of supports needed to increase the likelihood that a one-to-one initiative will blossom in a school. These include extensive professional development for teachers, technical support, and positive attitudes to technology-enhanced learning by teachers and school leadership. The existence of implementation and outcome studies, with many of the former and few of the latter, helps us understand the relative strengths and weaknesses of each. From the perspective of Zucker and McGhee's (2005) evaluation framework, implementation studies are best at helping us understand and improve the initial inputs, interactions, and intermediate outcomes, and they provide some limited evidence of ultimate outcomes. That is, these studies have highlighted benefits in terms of twenty-first century skills (Penuel, 2006; Rockman, 2003), but are not able, due to their design, to provide evidence of laptop-based improvements in student learning supported by rising achievement test data.

Outcome studies are best at clarifying the increments in student achievement accruing from well-implemented laptop programs, as well as the relative importance of specific independent and mediating variables on the dependent variable of most interest: student achievement. On the other hand, outcome studies provide less insight about various progressions and regressions that may have occurred in a laptop program's life cycle in a particular school or district setting. Ideally, evaluations might incorporate both implementation and outcome approaches. However, the cost of such evaluations and the pragmatic design challenges in implementing quasi-experimental studies in school make this scenario possible, although it is likely to occur infrequently.

Scaling up as the key policy question: Moore's Law?

The Internet is becoming the town square for the global village of tomorrow.

Bill Gates

Gordon Moore, cofounder of Intel, made the observation in 1965 that the number of transistors per square inch on integrated circuits had

Table 7.3   Evaluation questions from a scaling-up perspective

Dimension	Key questions	Evaluation design
Depth = deep change	To what extent have teachers' beliefs really changed? To what extent have social interaction patterns changed in the classroom? To what extent have teachers' principles of teaching changed and been reflected in relevant strategies?	Multiple time points to assess changes in teachers' beliefs and strategies as well as classroom social interaction patterns
Sustainability = success of an innovation over time	Does an initially successful innovation survive over 3, 5, or 7 years? If not, why not?	Policy demand and funding for longitudinal evaluation designs
Spread = spreading reforms to more classrooms and schools	To what extent have reform norms spread to other classrooms and schools?	Policy, procedures, and professional development that support new norms
Shift in ownership = from an external to internal authority	To what extent has the reform moved from an externally initiated idea to one owned internally by teachers and schools?	Changing from a focus on "buy in" to a "capacity to sustain reform in light of changing circumstances" view

actually doubled every year since the integrated circuit was invented. He then predicted that this trend would continue for the foreseeable future. However, even though the pace has actually slowed down somewhat, data density has doubled approximately every 18 months. This latter feature of computing is the current definition of Moore's law, and it appears to be holding true.

If we take Moore's law seriously in terms of the exponential growth of one-to-one computing as an educational phenomenon, then we must seriously address the question of scaling up. Here we are speaking not only in terms of successfully moving an innovation from one school to many—the conventional and limited meaning of scale—but also of scaling up in terms of depth, spread, sustainability, and ownership of an innovation (Coburn, 2003; see Table 7.3). That is, we agree with Coburn's criticism with what she views as the numbers fixation in framing what is meant by scale. Framing scale in this wider context is a "difference that makes a difference" (Bateson, 1979, p. 99) in that it can sharpen our analytic lens in terms of the ecology of scaling up one-to-one computing from small-scale pilots to system-wide reforms.

Much of what we know about one-to-one programs is based on many small-scale studies, even though large-scale studies' adoption of one-to-one computing is already under way around the world (e.g., the multimillion dollar investment in OLPC by developing countries with multilateral support from international agencies). However, we want to raise the question of knowledge about scaling up as a critically important question in relation to policy and research on one-to-one computing at present (Coburn, 2003; Dede et al., 2005).

Evaluating emerging one-to-one technologies: the changing face of human–computer interaction

> Computers are incredibly fast, accurate and stupid. Human beings are incredibly slow, inaccurate and brilliant. Together they are powerful beyond imagination.
>
> **Albert Einstein**

It is important in thinking about scaling up questions to consider the nature of emerging innovations, given the frequency of upgrades and decreasing obsolescence cycles that have, if anything, fuelled not abated the pace of one-to-one adoption in and out of school (see Table 7.4). The design features of specific one-to-one digital tools are important to understand from an evaluation perspective so that the affordances and constraints of various tools can be articulated, examined, and maximized. Furthermore, in attending to the affordances and constraints of current one-to-one tools, lessons learned can inform design of future tools. If we examine the affordances and features of some widely used tools, some of their limitations may also become apparent.

Going meta: models of change to strengthen utility and accuracy

Of the four meta-evaluation attributes that we addressed earlier in this chapter, utility and accuracy merit some further development in evaluations of one-to-one computing because, while evaluations of one-to-one computing, we think, do a good job in addressing the dynamics of utility and accuracy within the parameters of technological innovation, they often are less powerful in comprehensively analyzing the

Table 7.4   One-to-one innovations: some future directions in portability,
connectivity, functionality, capacity, and ease of use

Affordances	Features	Constraints	Future solutions
Portability	Small in physical size	Small screen size	Flexible film/screen display
Connectivity	Built-in or add on:WLANSIM card—mobile; Infrared; Bluetooth	Connectivity bandwidth	3 G Bluetooth v. 1.2
Functionality	E-mail client, "to do" list, calendarWeb browser, WAPFlash playerAudio player; Video player	Low memory	Expansion memory card; Increase internal RAM capacity
Capacity	Internal memory, essential applications already installed; Extra memory space to install other applications	Limit	Wireless connectivity to access online software
Ease of use	Iconized user-friendly interface	Ever changing OS	Generic OS on PDA, mobile phone, and smart phone

educational change dimension of such innovations. For example, many evaluations do address issues of educational change, but do not typically engage in sufficient examination of the educational change models out of which they operate.

Few studies explicitly identify a model of educational change; rather, they leave it implicit, often referring to supports and constraints without specifying their underlying assumptions, the exact manner in which high aspirations for educational technology innovation will be attained, and the role of technology in the process of wider educational reform (i.e., scaling up). Such gaps have often left initiatives and research in educational technology open to accusations that they are naïve about the wider social and political dynamics of educational reform (Cuban, 2001) and prone to fall back on additive models of change as one "important" variable after another is identified (Hoban, 2002).

Portraying practices: zooming in and zooming out

The research on e-learning over the last decade, as we have noted, has been heavily influenced by the move toward what has been termed "scientifically based educational research" (Lagemann, 2000; National Research Council, 2002). One feature of this powerful discourse and policy move is that experimentally framed studies of one-to-one computing, or indeed any other technological innovation, may be less sensitive to ecological context than we have emphasized in our work. As such, in attending to the immediate effects and variables, the more long-term effects and outcomes within particular ecological niches may become less apparent to researchers and evaluators.

While we strongly support the emphasis on rigor in recent debates on educational research, from the perspective of an ecological model of technological innovation, we question the implicit causal model in RCTs, which is at odds with accumulating evidence on the nature of educational change based on ecologically oriented research (Byrne, 1998; Carrier, Roggenhofer, Kuppers, & Blanchard, 2004; Cilliers, 1998; diSessa, 2000; Hoban, 2002; Zhao & Frank, 2003). Some examples—what we could call "surprises from practice"—help us understand the sometimes contradictory and unexpected outcomes of technological innovation in education.

Surprises from practice: unexpected outcomes

The expected or anticipated outcomes from one-to-one computing initiatives do not always transpire in actual practice. This observation alerts us to the way in which technological innovation plays a somewhat unpredictable role in ecological niches. To take one example, earlier in this book we noted how laptop mobility was reinterpreted to mean three quite different deployment models allied with different pedagogical possibilities. So-called mobile laptops, in fact, were distinguished more by how they came to be used like desktops (at least in the initiatives first 2 years) than personal, mobile anywhere, anytime learning tools (Conway, 2005). As such, the laptops were deployed in fixed, floating, or fostered modes, with fixed dominating, and the initiative's primary goals of mobile personal learning tool were not being attained—at least initially.

In the home setting, we have also noted how Facer, J. Furlong, R. Furlong, and Sutherland's (2003) ethnographic study of computer use identified three metaphors that captured very different home computer use by families. As such, the expectation that computers retain a fixed meaning is illusory and unexpected outcomes may in fact be predictable, or at least worth anticipating, as a normal part of one-to-one computing integration. The fact that, for example, one-to-one computing initiatives are undertaken, more often than not, with school-going adolescents may provide some unexpected outcomes for communication-oriented software/tools. For example, short message service (SMS) was originally intended as a tool for business people, but only when it was discovered that teenagers had taken over and reinvented SMS, a communication-oriented function, for their own communication purposes did designers seek to simplify and streamline the SMS function on mobile phones. In a similar fashion, careful attention in evaluation of one-to-one computing in schools may lead to what Norman calls "human-centered design" (1999).

As a further example, ninth grade students in the Maine laptop program reported a decrease in both work rate and quality, highlighting another unexpected project outcome. This was especially surprising as one of the key rationales for the introduction of one-to-one computing in education is its promise of increasing achievement and overall productivity just as computing has done so very markedly in the business world. These unexpected outcomes draw our attention to the need for careful development of "insider" perspectives on one-to-one computing initiatives from an ecological perspective that recognizes from the outset how different meanings may be attributed to one-to-one digital tools, at times resulting in unexpected and maybe even unintended consequences.

Conclusion

According to Facer et al. (2003), "Humans have always had an intimate association with devices and technologies they have created and no matter what the technology, contemporary commentators have predicted that the consequences will be socially transformative" (p. 223). It is the task of evaluation to assist in understanding the transformations and unravel their social and educational significance. Thus, for example, while *Sesame Street* was not a new technology when it appeared on the media landscape in the late 1960s, it was a new departure in the use of a relatively new technology (Lesser, 1974). It led to now almost 40 years of intensive research focused on understanding how best to teach its curriculum to children from low-income communities in particular (Fisch & Truglio, 2001).

Like one-to-one computing, *Sesame Street* was expected to bridge social inequities. Like one-to-one computing, *Sesame Street* was meant to transform children's experience of learning. While the early expectations related to transformation of young children's learning or decrease in social inequities were not attained, *Sesame Street* has made a significant contribution to the education of young children around the world, promoted debate about high-quality educational programming for children, and been adapted for various cultural settings. One-to-one computing is a much more diverse and potentially more powerful educational intervention, and in all likelihood its reach will be greater than that of *Sesame Street*. For this reason, if no other, it is important to understand the impact of various one-to-one technologies on children in school and in the home and the interaction of use in these two spheres on each other.

In many respects, the evaluation of laptop projects and other one-to-one computing initiatives is at an early stage. A number of points are central to this conclusion. First, we need to know more about the contexts within which laptops prove fruitful in promoting learning in different contexts. Much of current research is focused on reading and writing and comparatively little on other subject areas. Second, while there has been some research directed at understanding how one-to-one computing provides anywhere, anytime learning (a cross-context claim, assumption, or goal), we know very little, for example, about the outcomes, effect, or impact of laptop initiatives based on quasi-experimental or experimental designs (Penuel, 2006).

Nevertheless, we think that a mix of implementation and outcome studies in future evaluations is inevitable and desirable. That is, the changing discourse on educational research evident in the United States and the United Kingdom, in particular, is drawing attention to the research design of evaluations and research on all types of technological innovations. While we disagree with the overly narrow scope (what we see as their questionable causal-linear logic underpinning the design of RCTs), we are sympathetic with the motivation and quest inspiring those advocating RCTs. Randomized control trials seek to develop clearer links between dimensions of one-to-one innovations, processes, and measurable outcomes. Finally, large-scale adoption of one-to-one computing makes understanding this phenomenon an important issue in education—the triple helix of key stakeholders: government, industry, and education. In this context, in the final chapter of this book, we highlight key themes we have addressed throughout this book and identify some possible future directions in research, policy, and practice.

Final thoughts on laptops for children **8**

Science fiction does not remain fiction for long. And certainly not on the Internet.

Vinton Cerf

Introduction

Ubiquitous computing is becoming increasingly ubiquitous—from a few schools in Australia to a number of states, several school districts, and thousands of schools in the United States to country-wide pro-grams in developing countries through the OCPL project. The debate of what the best technology is for ubiquitous computing is still going on; nonetheless, a rapidly growing number of students worldwide are obtaining and using their digital pencils, varying from laptops, Tablet PCs, and simplified $100 laptops to personal digital assistants (PDAs), gameboards, cell phones, and many other options. The penetration of information and communication technologies (ICTs) into classrooms has created opportunities for ubiquitous computing to take root in schools and affect the established cultural practices. Schools in turn shape what ubiquitous technologies are used, in what ways, and for what purposes.

Few people would challenge the claim that technology is creeping into our lives inexorably. This book has focused on this phenomenon by examining the discourses, initiatives, and impact of the digital pencil phenomenon in the lives of K–12 students. At its most basic, this book examines laptops in education: their justification and use and impact on learning, learners, and identity. However, this book also can be read as a case study of the human–machine relationship in learning contexts because technology integration in essence is an ongoing process of continuous interactions between humans and technology. It is within this wider perspective that we offer a radical and powerful vision of the future of educational computing: educational technology co-evolution and co-adaptation for digital citizenship. That is, we propose a framework that locates educational technology within an ecosystem in which the processes of co-evolution and co-adaptation are the central explanatory processes. This vision situates the digital pencil phenomenon in the context of reshaping the school–society relationship in order to prepare students for digital citizenship in an era of globalization.

As such, we locate discussion of schools and technology as part of wider context where schools, technology, knowledge, and society are all being redefined through globalization and the emergence of a "network society" (Castells, 1997). Writing about the changing nature of competencies expected of K–12 graduates in a globalizing era, Gardner identifies an emerging consensus in twenty-first century competencies that reveals, despite "vast and gritty difference[s] that exist within and between nations … a surprising convergence in what is considered a reasonable pre-collegiate education in Tokyo or Tel Aviv, in Budapest or Boston" (Gardner, 2004, p. 238). These twenty-first century competencies are:

- understanding the global system
- capacity to think analytically and creatively across disciplines
- ability to tackle problems and issues that do not respect disciplinary boundaries
- knowledge and ability to interact civilly and productively with individuals from quite different cultural backgrounds
- knowledge and respect for one's own cultural traditions
- fostering of hybrid identities
- fostering of tolerance

The impact of changing relations among knowledge, schools, and the national and now global society in which we live is reflected in aims outlined by Gardner. It is our view that one-to-one learning is likely to play a critical role in helping education systems to attain these important educational aims.

When talking about human–machine relationships, we like to believe that we are the masters of machines. We invent them, use them, change them, and throw away old machines for new ones. Scientific fictions and movies such as *The Matrix* have warned us of the disastrous consequences of machines taking control of humans. From this point of view, technology in education is generally treated as the slave to teachers and students. Too often we ask what technology is good, what technology is bad, and how technology can be used to serve educational purposes.

However, this is only one side of the coin. Technology is not good, not bad, and not neutral—because the use of technology has impact on the users and the environment (Lei, 2005; Seidensticker, 2006; Zhao & Frank, 2003; Zhao & Conway, 2001). The human–machine relationship is not an either–or dilemma. Neither is the master or the slave of the other. To understand the relationship between technology and education, we must not only understand what we, as users, can do to technology, but also understand what technology can do to us and to the organization, the system, and the social context where we use technology. We like to think of technology as a living species that interacts with other species and the school ecosystem, as suggested by several researchers (Bruce & Hogan, 1998; Dede, 1996; Lei, 2005; Zhao & Frank, 2003).

The possibility of mobile learning (m-learning) on a large scale may support a more radical redefinition of schooling ecosystems as we know it. While we imagine schools as we know them will not disappear, there is likely to be a significant change in the next few decades as one-to-one technology enhanced learning contributes to what the OECD (2003) identified as one of three possible future scenarios (i.e., reschooling). We think it is unlikely that the status quo will prevail (one of the other OECD scenarios) or that deschooling (the third OECD scenario) is likely to present a viable or desirable educational future. However, we think one-to-one computing will make a significant contribution to reschooling in which a learner's identity and sense of agency is likely to shift significantly.

Thus, as we have discussed in earlier chapters, a school system can be viewed as an example of an ecosystem that is a combination of diverse components and various relationships (Zhao & Frank, 2003). In this complex social environment, various groups are closely connected with each other and form a network of changes. Technology in schools is not independent and isolated artifacts, but rather is situated in the network of changes and connected with the context. When a new technology is introduced into the system, it affects both informal and formal activities (Nardi & O'Day, 1999, p. 17), and it may also affect the social relationships of existing groups. Like species in the natural ecosystem, technology interacts with the users and the school system.

For example, the introduction of a new technology often requires the installation of new facilities, modification of existing policies or establishment of new policies and regulations, relocation of resources, reallocation and change in the informal and formal activities, and it may also affect the social relationships of different groups of people. When a technology is introduced into a school system, other things also change; these changes cause further changes in the technology use, which, in turn, affect how other things change. Changes caused by the interactions between technology and the school system not only determine how the technology is being used, but also affect teachers and students, the organization, and operation of the school system. For example, students might change their learning practice because of new technologies, and their learning practice might also influence how they use technology. Thus, both their technology use and learning practice evolve together.

Thus, changes in schools are bidirectional or even circular (Keiny, 2002), which is akin to the ecological process, in which genes, organisms, and the environment continuously interact with each other, co-evolve, and co-adapt to each other, and these interactions shape not only the organism but also the environment (Lewontin, 2000). In this final chapter, we return to this ecological framework to review the relationships among technology, teachers and students, and education. Then, within this framework, we will discuss the future of ubiquitous computing.

Human–machine relationships: teachers, students, and computers

Don't send a human to do a machine's job.

AI council member Smith, *The Matrix* (movie)

Based on the "machine-as-slave" mindset, our thinking of the human–computer relationship in schools has also focused on the control of teachers and students over computers. For example, the title of Robert Taylor's book, *The Computer in the School: Tutor, Tutee, Tool* (1980), reflects how the computer slave serves the human master: as a tutor to help students learn. By the way, this tutor does not decide what and how to teach; human masters do: as a tutee to be taught by students or as a tool to help students learn. We often neglect the fact that the development of technology changes not only the manner in which we interact with machines in educational settings (Anderson, 2006), but also the roles we play and the ways we work and entertain.

Enough has been said about how we as users select, adopt, adapt, and change technology, but little said about how technology changes us—which is happening in our daily lives. Here we want first to review a few of the latest technology trends and then discuss how the use of these technologies is changing us as the users.

Booming of Web 2.0

One trend is the booming of "Web 2.0." Although it is still a controversial concept, Web 2.0 is becoming an increasingly common phenomenon and attracting increasing amounts of attention. In general, Web 2.0 features collaboration, interactivity, user autonomy, and data control. The wide spread of social-networking Web sites, data-sharing Web sites, blogs, podcasting, and wikis is making the Internet "more important than ever, with exciting new applications and sites popping up with surprising regularity" (O'Reilly, 2005). Using these emerging technologies, young people today are building communities, creating media, and sharing their works (Stead, 2006).

One example is the thriving of social-networking Web sites. In 2005 and 2006, an online social revolution took place in many countries around the world, especially those in which children have relatively easy Internet access, with the launch and then exponential increase in membership of social-networking spaces such as Bebo and MySpace. It is estimated that today two out of every three people online in the United States visit social-networking sites such as MySpace and Facebook (Lamb, 2006). From "On the Internet, nobody knows you're a dog" to "On the Internet, *everybody* knows you're a dog" (Kinsley, 2006), increasing numbers of people are merging their real-world lives with their virtual ones through blogs, discussion forums, personal Web sites, and social-networking Web sites such as MySpace.

These new technologies have changed the roles of learners, who are not merely consumers of information and learning materials any more, but rather take on multiple roles, becoming "producers, collaborators, researchers and publishers" (Stead, 2006) A recent study conducted by Pew Internet & American Life Project found that 35% of online adults and 57% of teenagers age 12–17 create content and make their own content to post to the Web (Lenhart, 2006). Web 2.0 has been used for teaching and learning purposes. For example, some teachers use wiki Web pages as a venue for student collaboration on authentic learning tasks.[1] Blogging has been widely used not only by teachers to reflect their own teaching, voice their opinions on educational issues, and communicate with peers and friends, but also by students to reflect on their learning.

Penetration of ICT

Another trend is continuous penetration of all kinds of technologies. Some are also becoming increasingly ubiquitous. For example, since 2002, more than 67 million iPods have been purchased.[2] According to a recent survey conducted by the Pew Research Center, 20% of American adults and 26% of Internet users have an iPod or MP3 player, and about 12% of internet users say they have downloaded a podcast (Pew, 2006b). This report also points out the dramatic growth in the array of individuals and mainstream media institutions that now provide podcasts. One telling example is the remarkable expansion at Podcast Alley, a podcast directory Web site. The site listed fewer than 1,000 podcasts for download in November 2004 (Pew, 2006b) and has more than 27,000 different podcasts today; more than 1.1 million episodes are available for download,[3] and the number is growing quickly everyday. It is projected that by 2010, "12.3 million households will synchronize podcasts to their MP3 players" (Forrester, 2005).

The penetration of technology in our lives and the merging of the digital world and the real world are changing how we learn, work, entertain, and stay connected with family and friends in a world that is mediated by ICT all the time. But more changes are happening with the younger generation. Compared with their parents and teachers— the "digital immigrants"—they are more comfortable with technology because it has been a natural part of their lives. The average time this generation spends on all types of media every week is equivalent to a full-time job (Rideout, Foehr, & Roberts, 2005). Today, a typical 21-year-old in the United States has exchanged, on average, 250,000 e-mails, instant messages, and phone text messages, and spent 5,000 hours on game playing, 10,000 hours on cell phone use, and 3,500 hours online (Rainie, 2006).

> Ten years from now, teenagers are likely to enjoy a much richer panorama of options because the pursuit of intellectual achievement will not be tilted so much in favor of the bookworm, but

instead cater to a wider variety of cognitive styles, learning patterns, and expressive behaviors. (Negroponte, 1995, p. 220)

As Negroponte asserted rather presciently in 1995, digital learning would provide a new, more colorful and shared palette for self-expression and forging relationships—what he terms "e-xpression." The opportunities of the online social networking phenomenon are precisely what Negroponte anticipated 10 years ago in terms of unprecedented new contexts for children and adolescents to engage in e-xpression. For example, the online social networking revolution, mainly through the involvement of children and adolescents, points to the need to understand the dynamics of young people's online lives in an era moving toward ubiquitous computing. These dynamics involve not only patterns of media consumption (the conventional term and focus of research on children's media usage) but also patterns of production, representation, and expression.

Whether we look at the phenomenon of multitasking with media or use of multiple digital tools for self-expression, some valued and old but no less important educational questions need to be addressed. In terms of, for example, online social networking spaces, how do children make friends, learn, play safely, benefit or not in the short and long term from increased access to online social networks facilitated by one-to-one computing? Furthermore, what does balanced and all-round development of children and adolescents (typical and core aspirations in most countries curricular documents) mean in today's media-saturated world? Some key questions in relation to one-to-one computing are the following: How is one-to-one technology changing children's media consumption and production patterns? What is the role of adults and educators in mediating these patterns both in and out of school?

We may not know the answers to these questions at this point, but we know that their digital experiences have not only changed the ways the new generations communicate, socialize, and entertain, but also fundamentally changed how they approach learning. Children's out-of-school experiences with mobile, personal, and wireless technologies are beginning to create the m-generation who will inevitably challenge school's understanding of technologies' role in learning (Prensky, 2005/2006). The younger generation, accustomed to mobile and personal technologies, will not merely consume but also produce, not merely be shaped by technology but also shape its impact in their daily lives.

The penetration of ICT has, in fact, created a whole new generation—the digital generation, also called the N(et) generation (Tapscott, 1998), generation M(edia)/(ultitasker) (Rideout et al., 2005), or the digital natives (Prensky, 2001). This is the first generation growing up with digital technology, first computers, then the Internet and other ubiquitous information and communication devices such as game consoles, cell phones, PDAs, and iPods. Compared with their parents and teachers—the "digital immigrants"—they are innovative users of available technology and eager adopters of new technology, setting trends of technology use both in school and at home; They are using more kinds of technology, using increasingly sophisticated technology, at an increasingly earlier age, and using technology more regularly (NetDay, 2005). They are technologically savvy, confident in the positive value of technology, and reliant upon technology as an "essential and preferred component of every aspect of their lives" (NetDay 2005). They are multitaskers, often working on two or more tasks using two or more technology devices at the same time (Rideout et al., 2005).

To them, there is no clear distinction between play and learning. They have been learning from playing and playing while learning. They are not passive consumers of information, but energetic participants in and active contributors to the digital world. "Growing up digital," they are natural players in the digital world and they are shaping and creating the digital world. Technology has changed the way they entertain and learn, and they are pushing traditional learning to learning with technology to e-learning to m-learning.

From learning with technology to e-learning to m-learning: the co-evolution and co-adaptation of technology and education

We shape our buildings; thereafter they shape us.

Winston Churchill

Technology and learning have always shaped and been shaped by each other. On the one hand, how educational organizations function affects what technology survives in schools, what technology is selected out, and how technology evolves (Lei, 2005; Tan, Lei, Shi, & Zhao, 2003; Zhao & Frank, 2003). Television has taken root in most classrooms, but instructional film has never found a secure niche; whiteboards are replacing blackboards; and computer projectors are replacing overhead projectors. Many stories have illustrated how the survival and prosperity of a technology in schools is determined by its interactions with other components in the school system.

On the other hand, technology is playing a very active role in changing schools and education. For example, the book can be seen as one of the first mobile and personal learning tools. Even before that, the tablet and chalk provided (and still do in many developing countries) a mobile and personal learning tool for students. When everyone was able to afford a pencil, it changed how people learned (Papert, 1980). Similarly, when everyone regularly uses one or more personal computing devices, changes in learning can be expected (Chan et al., 2006).

We have observed the changes that are happening. For example, technology certainly has changed the physical aspects of traditional schools. Libraries are changed into computer labs and then media centers. Classrooms are wired and rewired (and unwired). Furniture is rearranged to allow for the mobility within the classroom. The instructor's computer is sitting next to the TV. More outlets are installed. After two decades of heavy investment in technology hardware, the outlook of schools has certainly changed.

But more importantly, technology has changed how schools function and how teaching and learning are conducted. Learning is extended outside formal class sessions and outside school, and becomes more intercurriculum (e.g., Twining et al., 2006). Students take their digital devices to real field trips or take virtual field trips using their digital technologies. They are sharing their work and collaborating on projects with peers in the same class, outside the school, or even from other countries. Organizations such as the World-links and the ORACLE education foundation are connecting students from the world. Learning is becoming increasingly diversified and globalized.

Moreover, new educational organizations such as virtual schools and virtual universities that facilitate new ways of teaching and learning are being created, adding new species to the education system. Virtual courses, online universities, and online learning systems are just a few examples. In fact, technology-facilitated distance learning became an indispensable component of postsecondary education long ago. As of 2000–2001, 90% of public 2-year and 89% of public 4-year institutions in the United States offered distance education courses using a variety of technologies, and had an estimated 3,077,000 enrollments in all distance education courses offered by 2- and 4-year institutions (NCES, 2003). According to Becta Research (Twining et al., 2006), in 2006 82% of colleges in the United Kingdom used a virtual learning environment. There are a plethora of data, reports, and studies on distance learning in postsecondary education. Here we want to focus more on the comparatively new development of online learning in K–12 settings.

Started and prospered in higher education, online learning has spread to K–12 education and is growing quickly. For K–12 students, using online learning systems, taking online courses, and attending online schools are becoming common practices. In Hong Kong, an online integrated learning environment (ILE) is used in some primary schools to cater to individual differences (Lee, 2006). In the United States, 57% of school districts host online instructional applications via a Web-hosting model (Quality Education Data, 2005). Schools employ all kinds of technology, such as streaming audio and video, computer animations, e-mail, newsgroups, chat rooms, bulletin boards, and digital portfolios. Students take virtual field trips, attend e-conferences, and learn with peers from other schools, states, and even other countries. Many organizations offer online learning opportunities for K–12 students. For example, NASA held e-conferencing on science programs including aviation history, planetary exploring, and weather tracking (McDermon, 2005).

The growth of virtual schools is even more impressive. The 2005 NetDay Speaking Out Event report finds that 17% of grade 6, 28% of grade 9, and 46% of grade 12 students say that they or someone they know has taken an online class (NetDay, 2005). The Virtual High School, one of the first precollege-level virtual schools, started offering Internet-based courses for the first time, in September 1997, to about 500 students in 27 schools in 10 states (U.S. Department of Education, 2000a). Less than 10 years later, in 2006, Virtual High School is offering 237 courses to 7,573 students in 394 U.S. schools located in 30 states and 25 international schools. Many states now have state-level virtual high schools, such as Michigan Virtual High School, Illinois Virtual High School, and Colorado Online Learning.[4] In Indiana, Indiana University offers more than 100 high school courses through its Independent Study Program.[5] In Hawaii, about 200–400 students coming from 30 to 48 secondary schools in the Hawaii Department of Education school system take online courses from the Hawaii E-School every semester.[6] In fact, online learning has become such an important component of the student learning experience that, in April 2006, the state of Michigan signed a law to require all high students to take at least one online course as part of high school graduatuation requirement (Michigan Virtual High School, 2006).[7]

In Canada, Fraser Valley Distance Education School, founded in 1990, has more than 500 full-time and 300 part-time online students.[8] In the United Kingdom, several virtual schools, such as the Satellite Virtual Schools,[9] Croydon Digital, Nisai Virtual Academy, and Briteschool, were set up for homeschoolers and students who cannot attend regular schools for various reasons. In Australia, Virtual School for the Gifted was established 10 years ago to cater to gifted students whose educational needs may not be met in regular classrooms.[10] Started in 2002, ENO-Environment Online is a global virtual school for sustainable development and environmental awareness, now serving 300 schools from 90 countries.[11] The e-school launched by the New Partnership for Africa's Development (NEPAD) aims to benefit all 600,000 elementary schools in Africa.[12]

After nearly two decades of development, virtual high schools have evolved into a wide spectrum of varieties. The courses offered vary from general K–12 curriculum to specific subjects such as science, mathematics, or foreign languages; the targeted audience varies from all students to targeted groups of students such as gifted students, students with special needs, and home-schoolers; the level of curriculum varies from general courses to advanced courses and college preparation courses. With expanded options and flexible scheduling, virtual schools are providing students with unprecedented learning opportunities.

The future of one-to-one computing

We can only understand life backwards; we must live life forwards.

Soren Kierkegaard

It is difficult and sometimes dangerous to predict the future (Seidensticker, 2006). One common mistake is that we tend to overestimate the short-term effect of technology, but underestimate the long-term effect, as Amara's law states. Thirty years ago, Ken Olson, founder of Digital Equipment Corp, stated, "There is no reason for any individuals to have a computer in their home" (Seidensticker, 2006, p. 23). This was true for most people about 15 years ago, with only 15% of U.S. households having computers in 1990 (U.S. Department of Labor, 1999). However, in 2006, according to Forrester Research (Schadler & Golvin, 2006), 90 million U.S. households had at least one personal computer (accounting for 80.4% of all households), and more than half (52.2%) of these households had multiple computers. Today, to many people, having a computer connected to the Internet is not only reasonable, but also indispensable.

However, we may not need to risk repeating similar mistakes when it comes to projecting the future of one-to-one computing, because its future is currently emerging, pushed by three major forces: One is the penetration of technology in our lives, the second is the digital generation's increasing demand of using mobile technology anytime, anywhere, and the third is the continuous investment in school technology with rising expectations about education's role in fostering the knowledge society. These three forces, working from bottom up and top down simultaneously, are pushing schools from N:1 computing to 1:1 computing to 1:N seamlessly ubiquitous computing.

The widespread adoption of technology

This book has listed plenty of numbers and examples on the rapid penetration of technology in our lives. Here we do not want to repeat ourselves, but just to add one more aspect of this penetration: We are also becoming increasingly connected. The Internet, global positioning system (GPS), radio frequency identification (RFID), Bluetooth, Instant Messaging—We are connected in various ways.

Take Internet connectivity as an example. In increasing numbers of places, wireless connectivity has expanded outside homes, offices, or organizations to the whole city or county. Many cities and regions in the world are building wireless networks for their citizens. In Australia, the largest wi-fi network—10,000 wireless access points across 1,700 locations—was built to provide connectivity to 540,000 students and 42,000 teachers (Gedda, 2006). In Taipei, a citywide network has over 4,000 hot spots, covering around 90% of the 2.6 million people in the city (Nystedt, 2006). According to a report on Muniwireless (muniwireless.com), a Web site focusing on citywide wireless projects worldwide, as of September 2006, 281 region or citywide networks in the United States were either in place or being planned, up from 122 in July 2005. In a recent study conducted by Pew Internet & American Life Project, a majority of people surveyed agreed that "a global, low-cost network will be thriving in 2020 and will be available to most people around the world at low cost" (Anderson & Rainie, 2006).

Modern information and communication technology not only is, as traditionally conceived, a new tool that we use to enhance our lives in the physical world, but also has created a whole new digital world. In this new world we use different technologies to seek and provide resources and information, express ourselves, communicate with others, create, consume, and entertain, often assuming new identities. The scope of the digital world is comparable to that of the real world. From e-learning, e-business, and e-governance to online gaming, online dating, and virtual marriage, activities happening in the digital world include everything imaginable.

The impact and the scope of these activities are not limited to the digital world; they are reaching our physical world. Interactions with ICT have already become an essential part of our daily lives. We learn, work, entertain, and stay connected with family and friends in a world that is mediated by ICT all the time. The digital world is penetrating into our physical world, becoming a part of our real life. There is no doubt that, in the future, our world will be further digitized. This is even more so, both now and in the future, for our children—the digital generation.

The growing demand of m-learning from the digital generation

The intertwined nature of childhood and technology is especially evident today, as the dynamics of children's and adolescents' digital lives change with the pace of technological innovation, particularly due to the undoubted appeal to them of contemporary mobile and personal technologies.

For those "growing up digital," laptop use has already become an important component of this generation's life from an early age. According to the National Report on NetDay's 2005 Speak Out Event, both the availability and sophistication of technology have increased significantly for today's students. The percentages of people reported using a laptop on a weekly basis for K–3 students, students grades 3–6, students grades 6–12, and teachers are 21, 28, 35, and 39%, respectively. Of K–3 students, 73% reported regular use of computers in their free time, and that percentage rose to 94% for students in grades 3–6. There was more than a 10% jump in the percentage of grade 6–12 students reporting using a number of ICT devices, including laptop computers, MP3 players, digital cameras, and video cameras (NetDay, 2005).

Few would question that in today's world children's out-of-school lives are characterized by increasing access to new mobile and personal digital technologies (Lenhart, Madden & Hiltin, 2005). Thus, it is natural to ask this question: "Will students who come to expect mobile, connected, personal devices outside school demand to use them within school?" (Chan et al., 2006)

The answer to this question is "yes." In fact, researchers predict that "schools will have difficulty keeping up with the range of personal devices that students bring to class" (Roschelle & Chan, 2005).

When students who participated in the NetDay Speaking Out Event were asked about the number 1 thing they would do if they were "designing a new school for students just like [them]," the most popular choice (33%) was "to provide laptops for every student that can be taken home." Their choice was strongly supported by their teachers; 48% of them believed that providing laptops for students had the potential to improve students' success in school (NetDay, 2005).

American students' wish for personal computing devices is echoed by students from other countries and regions. On November 2, 2005, the U.K. BBC Newsround Web site launched a project to collect technology ideas from children for the government and the prime minister ("PM asks kids for technology hints").[13] The Web site asked students to express their ideas for the future on how technology could work and how they think technology would affect their lives. Kids from all over the country volunteered their opinions. The first comment posted on this Web site was about laptops for everyone: "Our school is already quite technological. We have interactive whiteboards in almost every room and we e-mail our homework quite a lot. The next step up would be laptops for students as all the teachers already have them" (Laura, 14, student at Midhurst).

This idea was supported by several students. For example, 10-year-old William thought that "having laptops in school is a good idea because then pupils can save and delete work when they need to. I think we should be able to take them home as well." Similarly, 9-year-old Luke from Stoke wanted to have a laptop at home as well: "I think children should have wireless laptops at home because they can have homework sent to them and it does not waste paper and electricity."

Continuous investment in school technology: preparing for life in the networked knowledge society

It is very hopeful that these students' personal laptop dreams will come true in the near future because policy makers have always embraced new information and communication technologies with tremendous enthusiasm, despite past failures, and they have always been able to garner sufficient support for the technology agenda and been able to fund the technology initiatives (Zhao & Lei, forthcoming).

Written for a European readership, *Can We Learn Digitally?* (Apple, 2002) noted that governments had spent billions of euros in the preceding years on integrating ICTs in education. Clearly, governments and educators around the world have decided that human–computer interaction in schools is an important and worthwhile educational goal, given that most developed countries are in their third phase of planning how digital technologies will be used in K–12 education (Zhao, Lei, & Conway, 2006). Beyond nation-states, the drive to integrate computers in schools is pressing ahead within the context of an international multilateral agenda by transnational organizations such as the OECD, the U.N. (e.g., UN ICT Task Force), the World Bank, APEC, and the European Union, among others (UNESCO, 2003; Zhao et al., 2006). Furthermore, the advent of 3G-enabled m-learning/one-to-one computing has renewed interest in the potential of educational technology to transform education.

In fact, some of these students probably will get their personal laptops very soon. According to the 2005–2006 National Technology Assessment conducted by Quality Education Data (2006), 47% of the school districts surveyed expect to mount a major technology initiative or upgrade effort focused on desktops, laptops, and handhelds during the next 24 months.

It is quite safe to predict that a growing number of schools, states, and countries will join this N:1–1:1–1:N school technology stream. Gary Stager, a pioneer of laptops in education, points out that "many new laptop schools pretend they invented the idea and disregard the lessons of their predecessors" (Stager, 2006). It might be acceptable to pretend they invented the idea, but it would be very costly (in every sense) if the lessons of their predecessors were disregarded.

In the following section, based on our work and review of others' work, we will summarize what we know about one-to-one and ubiquitous computing, and make some recommendations:

Recommendations: what we know and what we need to do

Digital citizenship: a posthardware view of educational computing

Since the first group of computers was put into schools for teaching and learning purposes, the major effort in educational technology has been to increase technology access in education. This effort has been reflected in policy making, research, and practices. For example, a review on educational technology polices in the last 20 years finds out that maintaining high-quality and equitable access to technology over the long term is a central focus in all the educational technology policy documents, with some policies specifically discussing the importance of developing sustainability plans for technology funding (Culp, Honey, & Mandinach, 2003).

This effort has proven very successful. After two decades of heavy investment and strong advocacy, technology access in schools has increased staggeringly (Kleiner & Lewis, 2003; *Education Week*, 2006). Today the student–computer ratio in many countries (e.g., Canada, Norway, Australia, Singapore, New Zealand, France, Korea) has been decreased to smaller than 10:1 or even under 5:1 (Technology Counts, 2005). Schools in the United States have the most technology access in the world (Technology Counts, 2005). Technology has changed computers from rare innovations to daily appliances that have penetrated into school buildings and classrooms. Providing sufficient technology access to teachers and students—the goal for which policy makers, educators, researchers, and practitioners have strived for more than two decades—has been, or in some cases will soon be, achieved. Generally speaking, schools in many countries, especially the developed countries, have entered a "posthardware" era in terms of technology integration.

The digital world is becoming an increasingly important and natural part of our lives. Then, in this digital world, in which technology defines talents, what kinds of talents are needed? What kinds of skills will our students have to learn? Thus, schools need to teach our children to learn how to become competent citizens of this new world. Digital citizenship is the ability to live in the digital world productively and be a contributing member to society. It includes the following fundamental concepts:

- knowledge of the nature of the digital world, including a sound understanding of the differences and connections between the physical and virtual worlds, the nature of technology and how different media work together, the nature of online/virtual activities, the nature of the digital world as a constantly expanding and evolving global network of individual and collective participants; and the ability to tell fantasy from reality
- positive attitude toward the digital world, including an appreciation of the complexity and uncertainty of the digital world,

positive attitude toward technical problems, effective strategies to approach technical problems (knowing where and how to obtain assistance), and effective strategies to learn new ways of communication and information sharing

- ability to use different tools to participate and lead in the digital world, such as to participate and lead online communities; to entertain, learn, and work; and to obtain and share information
- ability to use different tools to express, create digital products, and create and manage online communities

In these circumstances, to what extent might one-to-one computing access provide an early window on citizenship for children and adolescents? What forms of online participation have been learned in social networking spaces such as Bebo, MySpace, and other such spaces? To what extent can these spaces be thought of as places for digital citizenship or for the development of civic understanding? How does online digital citizenship relate to and change "physical world" civic participation? For example, given that, in Europe (European Commission, 2001), 26% of young people aged 15–24 years viewed school as their most important form of social and political participation (4 and 8%, respectively, mentioned political parties and public bodies), what role can schools play in supporting digital citizenship through utilizing a range of digital tools in school? What, if any, is the school's role in mediating students' online social networking experiences in terms of developing digital citizenship?

With more sophisticated research tools at their disposal, research firms advising political parties can use perception analyzers as a tool in instant response research to track second by second potential voters reactions' to election candidates (see Luntz Research, www.luntz. com). Marketing specialists can identify a consumer's likely online or offline spending patterns based on only a handful of key pieces of information, such as income (on income of parents in terms of children and adolescents), gender, ZIP code/address, and hobbies. Whether it is consumption, voting patterns, or politicians' engagement with voters, the increasingly mobile and personal nature of technology is changing contexts for citizenship.

For others, children's out-of-school experiences with mobile, personal, and wireless technologies are beginning to create the m-generation who will inevitably challenge school's understanding of technologies' role in learning (Prensky, 2005/2006). For those more hopeful about technology creep, the m-generation, accustomed to mobile and personal technologies, will not merely consume but also produce, not merely be shaped by technology but also shape its impact in their daily lives. Consequently, children will be more engaged as both learners and citizens and increasingly be ready to take on the challenges of living in diverse and complex knowledge societies.

Schools can and many already do play an important role in guiding students for life as digital citizens. Whole-class-produced Web sites that engage students from other schools nearby, from across contested boundaries, and from the other side of the world provide new and important forms of social engagement for students in educational settings. For example, a cross-border digital learning project in Ireland has brought together students from the North and South of Ireland to share common cultural heritage, discuss and learn about cultural differences, and undertake joint learning projects (Austin et al., 2003).

Today's one-to-one computing enhanced learning environments occur in a different context, with much wider availability of digital tools outside school, more reliable and durable digital tools, and much greater attention to the curricular and school contexts of technological innovation. All of these supportive factors create a new context within which to study and understand technology in education. We summarize some important insights about what we know about educational reform in terms of four constituencies central to understanding the role of technology in schools:

- children: not all students are digital natives, but many are
- policy makers: accepting that the potential of technology to transform education has been overstated in the past, and reframing and connecting the role of one-to-one computer enhanced learning environments with wider system-level reforms
- school leaders and teachers: understanding that school context matters in terms of leadership and learning
- partners: leveraging the triple helix of government–community, school, and industry support

Goodness of fit

Returning to two sets of data we used earlier reiterates this point. First, large-scale systemic studies (e.g., the Impact 2 study in the United Kingdom [Pittard, 2003]; PISA-based study in 31 countries [Fuchs & Woessmann, 2003]) of the impact of computing on student attainment point to the moderate or neutral impact of computing on achievement. Other large-scale studies using nationally representative data sets (e.g.,

U.S. students' performance in the NAEP on the 2001 history assessment; see Wenglinsky, 2005, 2005/2006). Pittard's study points to a moderate impact on student attainment, given supportive conditions.

Similarly, Wenglinsky's NAEP-based studies in the United States point to how the optimal role for technology depends on students' grade level and subject/content area. For example, his 1998 study demonstrated that, for younger students, the quality of teacher-directed computer use—defined as a focus on higher order skills rather than drills on routine tasks—impacted student achievement in reading, mathematics, and science. For high-school students (2001 study), using computers for generic academic tasks (e.g., word processing, using computers for projects, creating charts and tables, and communication via e-mail and chat rooms) predicted higher student achievement than use for subject-specific tasks such as reading primary source documents. Wenglinsky's studies, in particular, point to what we might term "goodness of fit" between computing and curriculum in the classroom. That is, generic claims about the impact of computing need to be moderated by an understanding of how computers fit within specific curriculum and educational niches.

Taking these lessons from different contexts, the implication for current intensified uses of technology in one-to-one computing enhanced learning environments seems to be that specific curricular goals need to be addressed with technologies most appropriate in meeting these goals. This may in part depend on the age level of students, the complexity of the topic, and the affordances and constraints of one or more digital tools in one-to-one enhanced learning environments.

Careful planning of computer use to attain worthwhile and ambitious curriculum goals seems increasingly important, given that Fuchs and Woessmann's (2004) PISA-based study (175,000 students in 31 countries based on nationally representative samples)—the largest ever study of the impact of computer availability and use on achievement—indicated through use of multivariate analysis that "once we control extensively for family background and school characteristics, the relationship gets negative for home computers and insignificant for school computers" (p. 1). None of the studies by Wenglinsky (2005/2006), Fuchs and Woessmann (2004), or Pittard (2003) was undertaken in one-to-one computer enhanced learning environments. As such, in many ways we are not in a position to say with confidence that one-to-one computer enhanced learning environments will increase or decrease student achievement. However, the goodness of fit between one-to-one computing and specific curriculum goals seems a central consideration in planning for such learning environments.

Careful in setting goals

As children's out-of-school lives change in terms of media consumption and production, school systems are increasingly under pressure to consider what Warschauer calls the "whys," the "why nots," and the "how" of one-to-one computing (Warschauer, 2005/2006; see also Warschauer, 2006; Warschauer et al., 2004). Acknowledging the two significant trends underpinning m-learning, one-to-one laptop use, and wireless broadband, Warschauer identifies what not to expect of one-to-one computing. Based on case studies undertaken over 2 years on laptop programs in 10 schools (3 in Maine and 7 in California), Warschauer typically did not find evidence that laptops increased test scores, reformed troubled schools, or erased achievement gaps.

In terms of the whys, on the other hand, there was evidence, according to Warschauer, that laptop programs promote the type of learning characterized as twenty-first skills (i.e., "information literacy" skills such accessing, analyzing, and critiquing information), enhanced student motivation and engagement, supported teachers' integration of technology across the curriculum, and, with a school-wide focus on writing, increased the amount and quality of students' writing.

Taking Warschauer's why nots and whys together strongly suggests that school systems and individual schools ought to be careful in setting expectations as they plan strategically in terms of the adaptation of one-to-one technologies that are unquestionably reshaping students' out-of-school lives.

As educational technologies both proliferate and converge (in some ways; see the preface to this book), the actual and perceived relationships between increasingly ambitious educational goals for all students—not just an elite group of students as might have been the case 40 years ago—and these technologies raises the question of how best to create a fit between learning goals and technology use.

Conclusion: the social life of the digital pencil

Co-evolution and co-adaptation involve the very social life of technology and the technological life of very social human beings. Our ecological view of technology necessitates that we take seriously the social life of the digital pencil. The terms we associate with our new digital lives are imbued with social meaning. The "digital divide" reminds us of the social inequities in the distribution of technology—even if the binary nature of the term masks the complexities of that distribution. The "digital generation" reminds us that it is within a particular historical and cultural moment that we are experiencing a digital revolution—even if the essentialist nature of the term obscures the vastly different participation rates in various digital experiences within and between societies. The notion of "digital learning" alerts us to the new tools and new modes of semiotic production afforded to children and learners of all ages—even if the seductive nature of the term might dupe us into thinking that conventional learning experience is no longer useful when it is. Digital citizenship, in our view, must address the dynamics and challenges of the digital divide, digital learning, and the digital generation.

Notes

Preface

1. http://www.learningwithlaptops.org/
2. http://www.txtip.info/pages/2/index.htm
3. http://www.etsb.qc.ca/en/EnhancedLearningStrategy/default.shtm
4. WWP Website: http://www.prn.bc.ca/Wireless_Writing_Program.html
5. http://lft.ngfl.gov.uk/
6. http://www.axon.co.nz/info/Seven_thousand_primary_teachers_get_laptops.htm
7. http://www.novell.com/australia/stories/wesley_college.html.
8. http://www.pc-ap.fujitsu.com/news/casestudies/CGS%20Double.pdf

Chapter 1

1. Merriam–Webster Online, 2006.
2. http://www.britishmm.co.uk/brighton.asp
3. http://inventors.about.com/
4. http://www.bts.gov/publications/national_transportation_statistics/2005/html/table 04_09.html
5. Computer Take Back Web site, http://www.computertakeback.com/the_problem/poisonpcstoxictvs.cfm
6. http://www2.fht-esslingen.de/studentisches/Computer_Geschichte/grp4/marki.html
7. http://www.jmusheneaux.com/index52.htm
8. http://www.internetworldstats.com/stats.htm. Retrieved November 20, 2006.
9. http://www.internetworldstats.com/stats2.htm. Retrieved November 20, 2006.

10. comScore Networks, 2006. Retrieved November 20, 2006, from http://www.comscore.com/press/release.asp?press=1050
11. http://news.com.com/2061-11199_3-5950902.html
12. http://www.podcastingnews.com/archives/2006/04/podcasts_surpas.html

Chapter 2

1. *eSchool News Online*: http://www.eschoolnews.com/news/showStory.cfm?ArticleID=6060

Chapter 3

1. http://www.etsb.qc.ca/en/EnhancedLearningStrategy/Manuals.shtm
2. http://www.prn.bc.ca/iBookParentManual.htm

Chapter 4

1. Yechen Zhao is a coauthor of this chapter.
2. http://www.prn.bc.ca/wwp2005.pdf
3. http://news.bbc.co.uk/cbbcnews/hi/newsid_4390000/newsid_4399800/4399806.stm
4. Uncle Ben to Peter Parker in the movie *Spider-Man*.

Chapter 5

1. This chapter is coauthored with Nicole C. Ellefson and Kenneth A. Frank.
2. LeapPad is an electronic education toy resembling a talking book developed by LeapFrog Enterprises. It was awarded *Toy*

of the Year from Toy Manufacturers of America. Web site: www.leapfrog.com

3. A Pentop computer is a type of pen with a built-in computer processor inside that uses digital paper and pattern decoding technology to write, read, and speak. Web site: www.flypen.com

4. http://solutions.palm.com/regac/success_stories/SuccessStoryDctails.jsp?storyId=189

5. http://www.bbc.co.uk/schools/gcsebitesize/siteguide/index.shtml#website

6. http://edu.aweb.com.cn/2005/9/29/9130679.htm

7. http://www.epinions.com/cmd-review-2CAF-4B0B1EBC-3A5A29F9-prod3

8. E-school news online: http://www.eschoolnews.com/news/showstory.cfm?ArticleID=6740

9. http://www.bsf.gov.uk/

Chapter 7

1. http://www.bgfl.org/services/itaal/default.htm

Chapter 8

1. http://coolcatteacher.blogspot.com/2005/12/wiki-wiki-teaching-art-of-using-wiki.html

2. http://en.wikipedia.org/wiki/IPod, November 27, 2006.

3. http://www.podcastalley.com/, December 5, 2006.

4. Virtual Schools, http://www.shambles.net/pages/school/vschools/

5. Indiana University High School, http://www.indiana.edu/~scs/hs/highschoolcourses.html

6. http://www.eschool.k12.hi.us/Pages/some_answers.html

7. http://www.mivhs.com/upload_2/MIOnlineRequirment42106.pdf

8. http://www.fvdes.com/about/history.html

9. Satellite Virtual Schools, http://www.satellitevs.com/

10. Virtual School for the Gifted, http://www.vsg.edu.au/

11. NEO http://eno.joensuu.fi/basics/briefly.htm
12. NEPAD   http://www.nepad.org/2005/news/wmview.php?Art ID=30
13. http://news.bbc.co.uk/cbbcnews/hi/newsid_4390000/newsid_ 4399800/4399806

References

Agogino, A., Balacheff, N., Brecht, J., Brown, T., Chan, T. W., Dillenbourg, P., et al. (2006). Learning with the $100 laptop. Retrieved November 8, 2006, from http://www.g1to1.org/openletter.pdf

Anderson, J. Q., & Rainie, L. (2006). The future of the Internet II. Pew Internet & American Life Project. Retrieved November 15, 2006, from http://www.pewinternet.org/pdfs/PIP_Future_of_Internet_2006.pdf

Anderson, P. (2006). The future of human–computer interaction. In Becta Research, *Emerging technologies for learning.* Available online at http://partners.becta.org.uk/page_documents/research/emerging_technologies.pdf

Andrews, R., Freeman, A., Hou, D., McGuinn, N., Robinson, A., & Zhu, D. (2007). The effectiveness of information and communication technology on the learning of written English for 5- to 16-year-olds. *British Journal of Educational Technology, 38*(2) 325–336.

Andrews, R. (2004). *The impact of ICT on literacy education.* London: RoutledgeFalmer.

Annenberg Public Policy Center. (2000). Public policy, family rules and children's media use in the home. Press release. Retrieved March 1, 2006, from http://www.annenbergpublicpolicycenter.org/02_reports_releases/report_by_category.htm#mediaandchild

Apple Computers (2002). *Can We Learn Digitally?* Available online at http://a512.g.akamai.net/7/512/51/1fdb1568a5920/www.appl.com/uk/education/learn_digitally/pdf/learn_digitally.pdf

Apple Computers (2005). *Research: What it says about 1 to 1 learning.* Available online at www.ubiqcomputing.org/Apple_1-to-1_Research.pdf

Attewell, J. (2005). Mobile technologies and learning: A technology update and m-learning project summary LSDA. Retrieved October 25, 2006, from http://www.m-learning.org/docs/The%20m-learning%20project%20-%20technology%20update%20and%20project%20summary.pdf

Austin, R., Abbott, L., Mulkeen, A., & Metcalfe, N. (2003). Dissolving boundaries: cross-national co-operation through technology in education, *Curriculum Journal, 14*, 1, 55–84.

Ba, H., Tally, B., & Tsikalas, K. (2002). *Children's emerging digital literacies: Investigating home computing in low- and middle-income families.* Center for Children and Technology. Retrieved July 5, 2003, from www.edc.org/CCT/publications_report_summary.%20asp?numPubld=37

Bartels, F. (2000). Reflections on the RCDS Laptop Program after one year. Retrieved December 20, 2005, from http://www.learningwithlaptops.org/files/Laptop%20Program%20Reflections.pdf

Bartels, F. (2002). Reflections on the RCDS Laptop Program after three years. Retrieved December 20, 2005, from http://www.learningwithlaptops. org/files/3rd%20Year%20Laptop%20Prog.pdf

Bateson, G. (1979). *Mind and nature: A necessary unity.* New York: E. P. Dutton.

BBC News. (2006a). Christmas goes digital for many. Dec. 24, 2006. Retrieved January 8, 2007, from http://news.bbc.co.uk/1/hi/technology/6190543.stm.

BBC News. (2006b). March 10, 2006. Available online at http://news.bbc. co.uk/2/hi/technology/4794920.stm

Beamish, A. (1999). Approaches to community computing: Bringing technology to low-income groups. In D. Schö, B. Sanyal, & W. Mitchell (Eds.), *High technology and low-income communities* (pp. 349–370). Cambridge, MA: MIT Press.

Bean, T. W. (2000). Reading in the content areas: Social constructivist dimensions. In M. Kamil, P. B. Mosenthal, P. D. Pearson, & R. Barr (Eds.), *Handbook of reading research.* Mahwah, NJ: Lawrence Erlbaum.

Becker, H. J. (2001). *How are teachers using computers in instruction?* Paper presented at the Annual Meeting of the American Educational Research Association, Seattle.

Becta. (2006). The ICT and e-learning in FE survey: Key findings. Retrieved December 9, 2006, from http://www.becta.org.uk/corporate/publications/documents/ICT_in_FE%20key_findings_rev.pdf

Bereiter, C., & Scardamelia, M. (1987). *The psychology of written composition.* Mahweh, NJ: Lawrence Erlbaum.

Black, P. & Wiliam, D. (1998). Inside the black box: raising standards through classroom assessment, *Phi Delta Kappan.* Available online at http:// www.pdkintl.org/kappan/k_v80/k9810bla.htm

Bonifaz, A., & Zucker, A. (2004). *Lessons learned about providing laptops for all students.* Newton, MA: Education Development Center. Available at http://www.neirtec.org/laptop/LaptopLessonsRprt.pdf

Boone, J., & Forsythe, H. (2004). LeapFrog SchoolHouse company backgrounder (media fact sheet). LeapFrog Enterprises.

Borja, R. (2004) Technology Counts '04 Asia. *Education Week.* Retrieved December 9, 2006, from http://counts.edweek.org/sreports/tc04/article. cfm?slug=35asia.h23

Boston, W. (2003). Is 4G the future? *Time.* Available online at http://www. time.comtime/magazine/article/0,9171,493250,00.html

Bruce, B. (1993). Innovation and social change. In Bruce, B. C., Peytoon, J. K., & Batson, T. (Eds.), *Network-based classrooms: promises and realities.* New York: Cambridge University Press.

Bruce, B. C., & Hogan, M. P. (1998). The disappearance of technology: Toward an ecological model of literacy. In D. Reinking, M. C. McKenna, L. D. Labbo, & R. D. Kieffer (Eds.), *Handbook of literacy and technology.* Mahweh, NJ: Lawrence Erlbaum Associates.

Bruce, B. C., & Levin, J. A. (1997). Educational technology: Media for inquiry, communication, construction, and expression. *Journal of Educational Computing Research, 17*(1), 79–102.

Burbules, N., & Callister, T., Jr. (2000). *Watch IT: The promises and risks of new information technologies for education.* Boulder, CO: Westview Press.

Bureau of Transportation Statistics, National Transportation Statistics. (2005). http://www.bts.gov/publications/national_transportation_statistics

Byrom, E. (1998). *Factors influencing the effective use of technology for teaching and learning: Lessons learned from the SEIRTEC intensive site schools.* Greensboro, NC: SERVE, Inc.

Caldwell, B. J., & Spinks, J. M. (1988). *The self-managing school.* London: Falmer Press.

Calkins, L. (1996). *The art of teaching writing.* Portsmouth, NH: Heinemann.

Campbell, D. T. (1971). Methods for the experimenting society. Paper presented to the Eastern Psychological Association, New York City.

Castells, M. (1997). *The information age: Economy, society and culture. Volume II. The power of identity.* Oxford: Blackwell Publishers.

Castronova, E. (2001). *Virtual worlds: A first-hand account of market and society on the cyberian frontier.* CESifo Working Paper Series No. 618. http://www.bepress.com/cgi/viewcontent.cgi?article= 1008&context=giwp

CBS News. (2006). School cell phone ban causes uproar, May 12, 2006. Retrieved January 8, 2007, from http://www.cbsnews.com/stories/2006/05/12/national/main1616330.shtml

Cevenini, P. (2006). Creating a "21st-century school" for learning and working together. *e-school News Online.* Retrieved January 8, 2007, from http://www.eschoolnews.com/news/showStory.cfm?ArticleID=6622

Chan, T., Roschelle, J., Hsi, S., Sharples, M., Brown, T., Patton, C., et al. (2006). One-to-one technology-enhanced learning: An opportunity for global research collaboration. *Research and Practice in Technology-Enhanced Learning, 1*(1), 3–29. Retrieved November 26, 2006, from http://www.g1to1.org/Chan%20et%20al2006_One-to-one%20technology-enha.pdf

Choi, Y. B., Abbott, T. A., Arthur, M. A., and Hill, D. N. (2007). Toward a future wireless classroom paradigm, *International Journal of Innovation and Learning, 4*(1), 14–25.

Cilliers, P. (1998). *Complexity and postmodernism: Understanding complex systems.* London: Routledge.

Cisco. (2006). Australian schools reap benefits of wireless network, for much less than the cost of wired. Retrieved December 10, 2006, from http://www.cisco.com/en/US/products/ps6087/products_case_study-0900aecd80321a65.shtml

Clark, R. E. (1983). Reconsidering research on learning from media. *Review of Educational Research, 53*(4), 445–459.

Clute, E. (2000). Laptops for all at junior high. [On-line]. Available at www.post-gazette.com/regionstate/20000924laptops3.asp.

Cole, M. (1996). *Cultural psychology: A once and future discipline.* Cambridge, MA: Belknap Press of Harvard University Press.

Colella V. (2000). Participatory simulations: Building collaborative under-standing through immersive dynamic modeling. *Journal of the Learning Sciences, 9*(4), 471–500.

Coleman, J. S. (1988). Social capital in the creation of human capital, *American Journal of Sociology*, 94, 95–120.

Collis, B. A., Knezek, G. A., Lai, K., Miyashita, K. T., Pelgrum, W. J., Plomp, T., & Sakamoto, T. (1996). *Children and computers in school*. Mahwah, NJ: Lawrence Erlbaum Associates.

Computer Economics. (2005). The falling costs of mobile computing. Retrieved December 9, 2006, from http://www.computereconomics.com/article.cfm?id=1084

Conway, P. F. (2005). *Laptops initiative for students with dyslexia or other reading and writing difficulties: Early implementation evaluation report*. National Centre for Technology Education (NCTE), Ireland. Available online at www.ncte.ie

Conway, P. F., & Zhao, Y. (2003). From Luddites to gatekeepers to designers: Images of teachers in political documents. In Y. Zhao (Ed.), *What teachers should know about technology. Perspectives and practices*. Greenwich, CT: Information Age Press.

Crowley, M. (2005). Laptops that will save the world, *New York Times*, Available online at www.nytimes.com

Csete, J., Wong, Y. H., & Vogel, D. (2004). *Mobile devices in and out of the classroom*. Paper presented at the ED-MEDIA Conference, Lugano, Switzerland.

Csikszentmihalyi, M. (1991). *Flow: The psychology of optimal experience*. New York: Harper Perennial.

Cuban, L. (2001). *Oversold and underused: Computers in the classroom*. Cambridge, MA: Harvard University Press.

Cuban, L. (1999). The technology puzzle. *Education Week*, 4, August, *47,* 68. Available online at www.edweek.org/ew/vol-18/43cuban.h18

Cuban, L. (1993). Computers meet classroom: Classroom wins. *Teachers College Record*, 95, 2, 185–210.

Cuban, L. (2006). 1:1 Laptops transforming classrooms: Yeah, sure, *Teachers College Record*, Available online at http://www.tcrecord.org ID Number: 12818

Culp, K. M., Honey, M., & Mandinach, E. (2005). *A retrospective on twenty years of education technology policy, Journal of Educational Computing, 32*(3), 279–307.

Daly, T. G. (2006). *School culture and the mediation of a technological innovation*. Unpublished doctoral dissertation, University College Cork, Ireland.

Daly, T. G., & Conway. P. F. (2006). Micro-sultural contestation and ideo-logical settlement: towards a polycultural framing of "School Culture." Paper presented at the European Educational Research Conference, Dublin, Ireland, September, 2006.

Davis, D., Garas, N., Hopstock, P., Kellum, A., & Stephenson, T. (2005). *Henrico County Public Schools iBook survey report*. Arlington, VA: Development Associates, Inc.

Day, J. C., Janus, A., & Davis, J. (2005). *Computer and Internet use in the United States: 2003*. U.S. Census Bureau.

Dede, C. (1996). The evolution of learning devices: Smart objects, information infrastructures, and shared synthetic environments. In The future of networking technologies for learning. Retrieved January 14, 2006, from http://www.ed.gov/Technology/Futures/dede.html

Department of Education. (2001). Woods announces 2 million pound laptop initiative for students with dyslexia. Press release from Minister of Education December 18, 2001.

Department of Education. (1997). *Schools IT 2000: A policy framework for the new millennium*. Dublin. Retrieved March 1, 2006, from http://www.irlgov.ie/

Dell. (2003) PDAs in the classroom. http://gvc.gvsd.mb.ca/pda/early_class.htm

Digital Europe. (2003). Did you know what is the "thumb-generation"? Retrieved January 8, 2007, from http://www.digital-eu.org/didyouknow/default.asp?id=29

diSessa, A. (2000). *Changing minds: Computers, learning, and literacy*. Cambridge, MA: MIT Press.

Education Week. (2006). The Information Edge news release. http://www.edweek.org/media/ew/tc/2006/TC06_press.pdf

Epstein, J. L. (1985). Home and school connections in schools of the future: Implications of research on parent involvement, *Peabody Journal of Education, 62*, 2, 18–41.

Epstein, J. L. (2001). *School, family, and community partnerships: Preparing educators and improving schools*. Boulder, CO: Westview Press.

European Commission. (2006). ICTs in schools. Brussels: European commision: information society and media. Available online at http://ec.europa.eu/dgs/information_society/index_en.htm

European Commission. (2001). *Freeze-frame on europe's youth for a new impetus*, Brussels: European commission: directorate-general for education and culture.

Evans, J., & Lunt, I. (2002). Inclusive education: Are there limits? *European Journal of Special Needs Education, 17*(1), 1–14.

Facer, K., Furlong, J., Furlong, R., & Sutherland, R. (2003). *ScreenPlay: Children and computing in the home*. London: Routledge Falmer.

Facer, K., Stanton, D., Joiner, R., Reid, J., Hull, R., & Kirk, D. (2004). Savannah: A mobile gaming experience to support the development of children's understanding of animal behavior. *Journal of Computer Assisted Learning, 20*(6), 399–409.

Finn, C. E., Stevens, F. I., Stufflebeam, D. L., & Walberg, H. (1997). The New York City public schools integrated learning systems project: Evaluation and meta-evaluation. *International Journal of Educational Research, 27*(2), 159–174.

Fisch, S. M., & Truglio, R. T. (2001). *G is for growing: 30 years of research on children and* Sesame Street. Mahwah, NJ: Lawrence Erlbaum Associates.

Florida, R. (2002). *The rise of the creative class: And how it's transforming work, leisure, community and everyday life*. New York: Perseus Books Group.

Forrester Research. (2005). The future of digital audio. http://www.forrester.com/Research/Document/Excerpt/0,7211,36428,00.html

Frank, K. A., Zhao, Y., & Borman, K. (2004). Social capital and the diffusion of innovations within organizations: Application to the implementation of computer technology in schools. *Sociology of Education, 77,* 148–171.

Friedman, T. L. (2005) *The world is flat: A brief history of the twenty-first century*. New York: Farrar, Straus and Giroux.

Fuchs, T., & Woessmann, L. (2004). What accounts for international differences in student performance? A re-examination using PISA data. Retrieved January 19, 2007, from http://ideas.repec.org/p/ces/ceswps/_1235.html#provider

Fujitsu. (2006). Case study. Retrieved October 29, 2006, http://www.pc-ap.fujitsu.com/news/casestudies/CGS%20Double.pdf

Garden Valley Collegiate. (2002). PDA's in the classroom http://gvc.gvsd.mb/ca/pda/early_class.htm

Gardner, J., & Galnouli, D. (2004). Research into information and communications technology in education: Disciplined inquiries for telling stories better. *Technology, Pedagogy and Education, 13*(2), 147–161.

Gardner, H. (2004). How education changes: Consieration of history, science and values, In M. Suarez-Orozco and D. Qin-Hilliard (Eds.), *Globalization: culture and education in the new millennium*. Berkeley: University of California Press.

Gedda, R. (2006). World's largest Wi-Fi network uses Linux: Schools in Victoria get secure access. *Techworld,* March 14, 2006. Retrieved December 9, 2006, from http://www.techworld.com/security/features/index.cfm?featureid=2326&pagtype=samecatsamechan

Ghani, J. A., & Deshpande, S. P. (1994). Task characteristics and the experience of optimal flow in human–computer interaction. *The Journal of Psychology, 128*(4), 381–391.

Greaves, T. (2000, May). One-to-one computing tools for life (technology information). *T.H.E (Technological Horizons in Education),* 27(10), 54.

Guthrie, J. T., & Wigfield, A. (2000). Engagement and motivation in reading. In M. Kamil, P. B. Mosenthal, P. D. Pearson, & R. Barr. (Eds.), *Handbook of reading research*. Mahwah, NJ: Lawrence Erlbaum.

Haertel, G. D., & Means, B. (2003). *Evaluating educational technology: Effective research designs for improving learning*. New York: Teachers College Press.

Hargreaves, A. (2000). Four ages of professionalism and professional learning. *Teachers and Teaching, 6*(2), 151–182.

Harris, W., & Smith, L. (2004). Laptop use by seventh grade students with disabilities: Perceptions of special education teachers. Retrieved March, 2005, from http://libraries.maine.edu

Harrison C., Somekh, B., Lewin, C., & Mavers, D. (2003). The ImpaCT2 evaluation: Quantitative and qualitative findings on the relationship between ICT and school achievement of 60 schools in England. Paper presented at CAL 2003, Queens University, Belfast.

Hayes, J. R., & Flowers, L. S. (1980). Identifying the organization of writing processes. In L. W. Gregg & E. R. Steinberg (Eds.), *Cognitive processes in writing*. Hillsdale: NJ: Erlbaum.

Hoffman, D. L., Novak, T. P., & Duhachek, A. (2002). The influence of goal directed and experiential activities on online flow experiences. *Journal of Consumer Psychology, 13*.

Honey, M., & Culp, K. M. (2005). Critical issue: Using technology to improve student achievement. Retrieved March 15, 2006, from http://www.ncrel.org/sdrs/areas/issues/methods/technlgy/te800.htm

Hong Kong Case Study Report. Retrieved October 8, 2006 from http://sitesm2.org/sitesm2_search/docs/CN003_narrative.pdf

Hoppe, U. M. (2006). One-to-one technology-enhanced learning: An opportunity for global research collaboration. *Research and Practice in Technology Enhanced Learning, 1*(1), 3–29.

Houser, C., Thornton, P., & Kluge, D. (2002). Mobile learning: Cell phones and PDAs for education. Proceedings of the International Conference on Computers in Education (ICCE'02).

Hubbard, L., Mehan, H., & Stein, M. K. (2006). *Reform as learning: School reform, organizational culture, and community politics in San Diego*. New York: Routledge.

Internet Usage Statistics—The Big Picture. World Internet users and population stats. Retrieved January 15, 2006, from http://www.internetworldstats.com/stats.htm

Jaillet, A. (2004). What is happening with portable computers in schools? *Journal of Science Education and Technology, 13*(1), 115–128.

Jaworski, B., & Phillips, D. (1999). *Comparing standards internationally: Research and practice in mathematics and beyond*. Oxford: Symposium Books.

Jeroski, S. (2003). *Wireless Writing Project School District No. 60 (Peace River North) research report: Phase II*. Vancouver, BC: Horizon Research and Evaluation, Inc.

Jeroski, S. (2005). Research report: The Wireless Writing Program 2004–2005. http://www.prn.bc.ca/wwp2005.pdf

Johnstone, B (2003a). Schools transformed, October 7. Retrieved March, 18, 2006, from http://www.theage.com.au/articles/2003/10/06/1065292519294.html

Johnstone, B. (2003b). *Never mind the laptops: Kids, computers, and the transformation of learning*. iUniverse Inc: Lincoln, Nebraska.

Joiner, W. (2005). Generation speed. *Seventeen Magazine,* November, 147–149.

Kamil, M. L., Intrator, S. M., & Kim, H. S. (2000). The effects of other technologies on literacy and literacy learning, In M. Kamil, P. B. Mosenthal, P. D. Pearson, & R. Barr. (Eds.), *Handbook of reading research*. Mahwah, NJ: Lawrence Erlbaum.

Keiny, S. (2002). *Ecological thinking: A new approach to educational change.* Lanham, MD: University Press of America.

Keefe, D., Farag, P., & Zucker, A. (2003). Annotated bibliography of ubiquitous computing evaluations. Retrieved 6/4/2007 from http://www.ubiq-computing.org/Reference.pdf

Kerr, K. A., Pane, J. F., & Barney, H. (2003). Quaker Valley Digital School District: Early effects and plans for future evaluation (technical report TR-107-EDU). Santa Monica, CA: RAND Corporation. http://www.rand.org/pubs/technical_reports/2004/RAND_TR107.pdf

Kinsley, M. (2006). Like I care: On the Internet, everybody knows you're a dog. *Slate,* November 27, 2006, http://www.slate.com/id/2154507

Kline, S., & Botterill. J. (2001). *Media use audit for B.C. teens: Key findings.* Report prepared for distribution to BC schools, Media Analysis Laboratory, Simon Fraser University: British Columbia. URL: http://www.sfu.ca/media-lab/

Kleiner, A., & Lewis, L. (2003). Internet access in U.S. public schools and classrooms: 1994–2002. Washington, DC: U.S. Department of Education. National Center for Education Statistics. 2003. Available online at http://nces.ed.gov/pubsearch/pubsinfo.asp?pubid+2004011

Klopfer, E., & Yoon, S. (2005). Developing games and simulations for today and tomorrow's tech-savvy youth. *Tech Trends, 49*(3), 33–41.

Klopfer, E. S., Yoon, S., & Rivas, L. (2004). Comparative analysis of palm and wearable computers for participatory simulations. *Journal of Computer Assisted Learning, 20*(55), 347–359.

Knowledge Networks/Statistical Research. (2002). More kids say Internet is the medium they can't live without. http://www.statisticalresearch.com/press/pr040402.htm

Kosma, R. (1994). Will media influence learning? Reframing the debate. *Educational Technology, Research and Development, 42*(2), 7–19

Krendl, K., & Lieberman, D. (1988). Computers and learning: A review of recent research. *Journal of Educational Computing Research, 4*(4), 367–89.

Lamb, P. (2006). Have YourSpace call MySpace. *The Christian Science Monitor,* commentary. Retrieved December 5, 2006, from http://www.csmonitor.com/2006/1108/p09s02-coop.html

Lankshear, C., & Bigum, C. (1999). Literacies and new technologies in school settings. *Curriculum Studies, 7*(3), 445–465. Available on-line at http://users.wantree.com.au/~peterh/rhetoric/

Lee, J., Luchini, K., Michael, B., Norris, C., & Soloway, E. (2004). More than just fun and games: Assessing the value of educational video games in the classroom. Proceedings of CHI 2004 Connect: Conference on Human Factors in Computing Systems, Vienna, Austria.

Lee, K. (2006). Online learning in primary schools: Designing for school culture change. *Educational Media International, 43*(2), 91–106.

Lei, J., & Zhao, Y. (2005). *Conditions for technology use in schools.* Paper presented at American Educational Research Association Annual Meeting 2005, Montreal, Canada. Also presented at Global Chinese Conference on Computers in Education (GCCCE) 2005, Hawaii, June.

Lei, J., & Zhao, Y. (2006). *What impact do different technology uses have on student outcomes?* Paper presented at American Educational Research Association Annual Meeting 2006, San Francisco, April 7–11.

Lei, J., & Zhao, Y. (2005). Computer uses and student achievement: A longitudinal study. *Computers & Education, 49*(2), 284–296.

Leo, P. (2005). Cell phone statistics that may surprise you, *Pittsburgh Post Gazette*, March 16, 2006, http://www.eng.vt.edu/pdf/upload_file/Cell%20phone%20statistics.pdf

Lemelson–MIT. (2004). Cell phone edges alarm clock as most hated invention, yet one we cannot live without. Retrieved January 14, 2006, from http://web.mit.edu/invent/n-pressreleases/n-press-04index.html

Lemke, C., & Martin, C. (2004). One-to-one computing in Indiana—A state profile (preliminary report). Retrieved February 6, 2006, from http://www.metiri.com/NSF-Study/INProfile.pdf

Lenhart, A. (2006). User-generated content. Pew Internet and American Life Project, http://www.pewinternet.org/PPF/r/76/presentation_display.asp

Lenhart, A., & Madden, M. (2007). *Social networking websites and teens: An overview.* Pew Internet and American Life Project. Retrieved January 12, 2007, http://www.pewinternet.org/pdfs/PIP_SNS_Data_Memo_Jan_2007.pdf

Lenhart, A., Madden, M., & Hitlin, P. (2005). Teens and technology: Youth are leading the transition to a fully wired and mobile nation. Pew Internet and American Life Project: Washington, DC. Available online at: http://www.pewinternet.org/pdfs/PIP_Teens_Tech_July2005web.pdf

Leo, P. (2005). Cell phone statistics that may surprise you. *Pittsburgh Post Gazette,* March 16, 2006, http://www.eng.vt.edu/pdf/upload_files/Cell%20phone%20statistics.pdf

Levin, H. M., & McEwan, P. J. (2001). *Cost-effectiveness analysis* (2nd ed.). Thousand Oaks, CA: Sage.

Levinson, P. (1998). *The soft edge: A natural history and future of the information revolution.* New York: Routledge.

Levy, S. (2004). No net? We'd rather go without food. *Newsweek,* October 11, 2004, 14.

Lewontin, R. (2000). *The triple helix: Gene, organism, and environment.* Cambridge, MA: Harvard University Press.

Liebert R. M., & Sprafkin, J. N. (1988). *The early window: Effects of television on children and youth* (3rd ed.). New York: Pergamon Press.

Lin, N. (2001). *Social capital: A theory of social structure and action.* New York: Cambridge University Press.

Lipsey, M. W., & Cordray, D. S. (2000). Evaluation methods for social intervention. *Annual Reviews Psychology, 51,* 345–375.

Luehrmann, A. (2002). "Should the computer teach the student ..."—30 years later. *Contemporary Issues in Technology and Teacher Education* [Online serial], *2*(3). Available: http://www.citejournal.org/vol2/iss3/seminal/article2.cfm

Madden, M. (2006). Internet penetration report. Retrieved November 20, 2006, http://www.pewinternet.org/pdfs/PIP_Internet_Impact.pdf

McCarthy, J., Wright, P., Wallace, J., & Dearden, A. (2005). The experience of enchantment in human–computer interaction. *Personal and Ubiquitous Computing*. Retrieved, January 10, 2006, http://www.personal-ubicomp.com/

McDermon, L. (2005). Distance learning: It's elementary! *Learning and Leading with Technology, 33*(4), 28–34.

McFarlane, A. (1997). What are we and how did we get here? In McFarlane, A. (Ed.), *Information technology and authentic learning: Realizing the potential of computers in the primary classroom*. New York: Routledge.

Means, B., & Penuel, W. R. (2005). Research to support scaling up technology-based educational innovations. In C. Dede, J. P. Honan, & L. C. Peters (Eds.), *Scaling up success: Lessons from technology-based educational improvement* (pp. 176–197). San Francisco: Jossey–Bass.

Mertens, D. M. (1998). *Research methods in education and psychology: Integrating diversity with qualitative and quantitative approaches.* Thousand Oaks, CA: Sage.

Microsoft Corporation. (1999). Anytime, Anywhere Learning: A guide to getting started. [On-line]. Available at http://www.microsoft.com/education/k12/aal

Microsoft Corporation. (2006). i-Safe and Microsoft bring Net safety to schools and parents. Creation of i-LEARN Online provides easy access to i-Safe's celebrated lessons. Available online at http://www.microsoft.com/presspass/press/2006/feb06/02-02iSAFEMSPR.mspx

Miles, K. H., & Darling-Hammond, L. (1998). Rethinking the allocation of teaching resources: Some lessons from high-performing schools. *Educational Evaluation and Policy Analysis, 20*(1), 9–29.

MIT Media Lab. (2006). Frequently asked questions one laptop per child. Accessed March 15, 2006, http://laptop/media.mit.edu/

MOBIlearn, (2005). Guidelines for learning/teaching/tutoring in a mobile environment. MOBIlearn/UoN, UoB,OU/WP4/D4.1/1.2.

Murphy, R. F., Penuel, W. R., Means, B., Korbak, C., Whaley, A., & Allen, J. E. (2002). E-DESK: A review of recent evidence on the effectiveness of discrete educational software. SRI for U.S. Dept. of Education. SRI Project 11063.

Murray, C. (2006). "School of the future" opens doors: First-of-its-kind school seeks to meet the needs of 21st-century learners. *eschool news online,* http://www.eschoolnews.com/news/showStory.cfm?ArticleID=6579%20%20

Naismith, L., Lonsdale, P., Vavoula, G., & Sharples, M. (2005). Literature review in mobile technologies and learning. Report for NESTA Futurelab, Report 11. Available online at http://www.futurelab.org.uk/research/reviews/reviews_11_and12/11_01.htm

Nardi, B. A., & O'Day, V. L. (1999). *Information ecologies*. Cambridge, MA: The MIT Press.

National Center for Education Statistics (NCES). (2003). *Report of distance education at degree granting postsecondary institutions: 2000–2001*. Washington, DC: U.S.12/05/2007 http://nces.ed.gov/pubs2003/2003017.pdf

National Center for Education Statistics (NCES). (2004). *Digest of education statistics, 2004*. Retrieved January 29 from http://nces.ed.gov/programs/digest/d04/

National Transportation Statistics, 2005 http://www.bts.gov/publications/national_transportation_statistics/2005/html/table 04_09.html

Naughton, J. (2000). *A brief history of the future*. London: Orion Publishing Group.

NCREL. (2000). The road less traveled: Restructuring budgets to accommodate hybrid expenditures. *Policy Issues, February,* 4. Retrieved February 26, 2006, http://www.ncrel.org/policy/pubs/html/pivol4/pivol4_2.htm

Negroponte, N. (1995). *Being digital*. New York: Vintage Books.

NetDay. (2005). *Our voice, our future: Student and teacher views on science, technology and education.* https://www.nacs.org/whitepapers/NETDAYSpeakUpReport.pdf

Newhouse, C. P. (2001). A follow-up study of students using portable computers at a secondary school. *British Journal of Educational Technology, 32*(2), 209–219.

Newhouse, C. P., & Rennie, L. (2001). A longitudinal study of the use of student-owned portable computers in a secondary school. *Computers and Education, 36*(3), 223–243.

Newsweek. (2002). Put yourself here. November 25, 2002, 60.

Newsweek. (2005). Data points. June 27, 2005, E2.

Norman, D. A. (1999). *The invisible computer: Why good products can fail, the personal computer is so complex, and information appliances are the solution.* Cambridge: MIT Press.

Nussbaum, M., & Zurita, G. (2005). A conceptual framework based on activity theory for mobile CSCL. *British Journal of Educational Technology, 38*(2), 211–235.

Nystedt, D. (2006). Taipei's city-wide Wi-Fi passes test. *InfoWorld,* June 27, 2006. http://www.infoworld.com/article/06/06/27/79664_ HNtaipeiwifi_1.html

O'Donoghue, T. A., & Dimmock, C. A. J. (1998). *School restructuring: International perspectives*. London: Kogan Page.

OECD. (2003). *The PISA 2003 assessment framework—Mathematics, reading, science and problem solving knowledge and skills.* Paris: OECD.

Ofcom. (2006). Communications market reports 2006. Retrieved December 5, 2006, from http://ofcom.org.uk/research/cm/

OfSTED. (2000). Inspection report: The Cornwallis School. Available at http://www.ofsted.gov.uk/reports/118/118874.pdf

Oppenheimer, T. (1997). The computer delusion, *The Atlantic Monthly*, 280(1), 45–62.

O'Reilly, T. (2005). What is Web 2.0: Design patterns and business models for the next generation of software. http://www.oreillynet.com/pub/a/oreilly/tim/news/2005/09/30/what-is-web-20.html

Papert, S. (1980). *Mindstorms: Children, computers, and powerful ideas*. New York: Basic Books.

Papert, S. (1992). *The children's machine: Rethinking school in the age of the computer*. New York: Basic Books.

Papert, S. (1996). Computers in the classroom: Agents of change. *Washington Post Education Review*. Retrieved May 21, 2006, from the World Wide Web: http://www.papert.org/articles/ComputersInClassroom.html

Parsad, B., & Jones, J. (2006) Internet access in U.S. public schools and classrooms: 1994–2003. *Education Statistics Quarterly, 7*(1–2). http://nces.ed.gov/programs/quarterly/vol_7/1_2/4_3.asp

Pea, R. (2000). *The Jossey–Bass reader on technology and learning*. San Francisco: Jossey–Bass.

Penuel, B., Tatar, D., & Roschelle, J. (2004). The role of research on contexts of teaching practice in informing the design of handheld learning technologies. *Journal of Educational Computing Research, 30*(4), 353–370.

Penuel, W. R. (2005). *Research: What it says about 1-to-1 learning*. Cupertino, CA: Apple Computer, Inc. Available online at http://www.ubiq-computing.org/Apple_1-to-1_Research.pdf

Penuel, W. R. (2006). Implementation and effects of one-to-one computing initiatives: A research synthesis. *Journal of Research on Technology in Education*, 38, 329–348.

Penuel, W. R., Kim, D. T., Michalchik, V., Lewis, S., Means, B., Murphy, R., et al. (2002). *Using technology to enhance connections between home and school: A research synthesis*. Planning and Evaluation Service, U.S. Department of Education, DHHS Contract #282-00-008-Task 1. Retrieved March 4, 2006, from http://www.sri.com/policy/ctl/html/synthesis1.html

Perelman, L. J. (1996). Opportunity cost: How a deep-seated corporate obsession with the education market baked Apple. *Wired, 4,* 11, Retrieved February 28, 2006, http://www.wired.com/wired/archive/4.11/es_apple.html

Perry, D. (2003). Wireless networking in schools: A decision making guide for school leaders. Specialist Schools Trust: London. Also available online at http://www.becta.org/uk/page_documents/leas/wire.pdf

Pew Research. (2006). On podcasting. http://www.pewinternet.org/pdfs/PIP_Podcasting.pdf

Picus, L. (1997). The challenges facing school districts in budgeting for technology. Available at http://www.bellsouthcorp.com/bsf/technology/picus/htm

Pitler, H., Flynn, K., & Gaddy, B. (2004). *Is a laptop initiative in your future?* MCREL Policy Brief. Available at http://www.mcrel.org/PDF/PolicyBriefs/5042PI_PBLaptopInitiative.pdf

Pittard, V. (2003). ImpaCT2: The impact of information and communication technologies on pupil learning and attainment. Becta Research. Retrieved December 1, 2006, from http://www.becta.org.uk/publications

Portes, A. (1998). Social capital. Its origins and applications in modern sociology. *Annual Review of Sociology, 24,* 1–24.

Prensky, M. (2005/2006). Listen to the natives. *Educational Leadership, 63*(4), 8–13.

Prensky, M. (2001). Digital natives, digital immigrants. *On the Horizon, 9*(5), 1–2.

Putnam, R. D. (1993). *Making democracy work. Civic traditions in modern Italy,* Princeton University Press: Princeton.

Quality Education Data (QED). (2004). 2003–2004 Technology purchasing forecast. http://www.qeddata.com/marketkno/researchreports/techpurchaseforecast.aspx

Quality Education Data (QED). (2006). 2005–2006 National technology assessment. http://www.qeddata.com/MarketKno/ResearchReports/nta.aspx

Quayle, E. (2004). Assessment issues with young people who engage in problematic sexual behavior through the Internet. In M. C. Calder (Ed.), *New developments with young people who sexually abuse.* Lyme Regis: Russell House Publishing.

Quayle, E., & Taylor, M. (1999). Young people who sexually abuse: The role of the new technologies. In M. Erooga & H. Masson (Eds.), *Children and young people who sexually abuse others.* London: Routledge.

Rainie, L., & Keeter, S. (2006). Pew Internet Project Data memo, retrieved May 17, 2007, http://www.pewinternet.org/pdfs/PIP_Cell_phone_study.pdf

Reuters. (2006). Young favor internet over TV. Retrieved 12/05/2006, http://news.yahoo.com/s/nm/20061129/wr_nm/internet_television_dc

Raudenbush, S., & Bryk, A. (2002). *Hierarchical linear models* (2nd ed.). Thousand Oaks, CA: Sage.

Rideout, V. J., Foehr, U. G., & Roberts, D. F. (2005). *Generation M: Media in the lives of 8- to 18-year-olds.* Washington, DC: A Kaiser Family Foundation Report. Available online at: http://www.kff.org/entmedia/index.cfm

Rideout, V. J., Ula, G., Foehr, U. G., Roberts, D. F., & Brodie, M. F. (1999). *Kids and media @ the new millennium: A comprehensive national analysis of children's media use.* Washington, DC: A Kaiser Family Foundation Report. Available online at http://www.kff.org/entmedia/index.cfm

Robinett, C., Leight, M., Malinowski, C., & Butler, J. (2005). *K–12 One-to-one computing handbook.* Center for Digital Education. Available online at http://www.centerdigitaled.com/

Rockman, S. et al. (2003). Learning from laptops. *Threshold,* Fall, 24–28.

Rockman, S. et al. (1997). The laptop research study: First year study of the laptop program [On-line]. Available at http://www.microsoft.com/education/k12/aal

Rockman, S. et al. (1998). Powerful tools for schooling: Second year study of the laptop program. San Francisco, CA. [Online]. Available at http://www.microsoft.com/education/k12/aal

Rockman, S. et al. (1999). The laptop research study: Third year study of the laptop program [On-line]. Available at http://www.microsoft.com/education/k12/aal

Rockman, S. et al. (2000). A more complex picture: Laptop use and impact in the context of changing home and school access. The third in a series of research studies on Microsoft's Anytime, Anywhere Learning Program. Retrieved May 16, 2006, from http://cdgenp01.csd.toshiba.com/content/publicsector/edgov/year_three_rockman_study.pdf

Rockman et al. (2006). The laptop research study: Third year study of the laptop program. Available online at http://www.microsoft.com/education/k12/aa

Rodríguez-Campos, L. (2004). Metaevaluation: Theoretical and practical guidelines for evaluating evaluations. *International Journal of Learning, 11.*

Rogers, E. M. (1995). *Diffusion of innovation.* New York: Free Press.

Romig, N., Yan, B., & Zhao, Y. (2004). *Inexpensive, interactive learning systems and their effects on reading achievement.* Available online at http://www.msu.edu~boyan/doc/leappad.pdf

Roschelle, J., & Chan, T-W. (2005). G1:1 envisions future scenarios of collaborative learning with technology: A brief summary of 2005 G1:1 workshop. Retrieved October 8, 2006, from http://www.g1to1.org/resources/efsclt_2k5.php

Roschelle, J., Pea, P., Hoadley, C., Gordin, D., & Means, B. (2000). Changing how and what children learn in school with computer-based technologies. *The Future of Children: Children and Computer Technology, 10*(2). Retrieved March, 10, 2006, http://www.futureofchildren.org/pubs-info2825/pubs-info_show.htm?doc_id=69787

Rowan, L., Knobel, M., Bigum, C., & Lankshear, C. (2002). *Boys, literacies and schooling: The dangerous territories of gender-based literacy reform.* Milton Keynes, England: Open University Press.

Russell, M., Bebell, D., & Higgins, J. (2004). Laptop learning: A comparison of teaching and learning in upper elementary classrooms equipped with shared carts of laptops and permanent 1:1 laptops. Boston: Technology and Assessment Study Collaborative, Boston College.

Samuelson, R. J. (2004). "A cell phone? Never for me." *Newsweek,* August 23, 2004, p. 63.

Sandholtz, J. H., Ringstaff, C., & Dwyer, C. D. (1997). *Teaching with technology: Creating pupil centered classrooms.* New York: Teachers College Press.

Sarason, S. (1991). *The predictable failure of educational reform.* New York: Teachers College Press.

Schadler, T., & Golvin, C. (2006). The state of consumers and technology: Benchmark 2006, North American consumer technology adoption study. Available at http://www.forrester.com/Research?Document?Excerpt/0,7211,38868,00.html

Schaumburg, H. (2001). The impact of mobile computers in the classroom: Results from an ongoing video study. Center for Media Research, Freie Universitaet Berlin. Retrieved March 5, 2004, from www.cmr.fu-berlin. de/~heike/conferences/aect01/aect01.pdf

Schwartz, J. (1993). Caution: Children at play on information superhighway: Access to adult networks holds hazards. *Washington Post*, p. 1. Retrieved March 5, 2006, from http://www.eff.org/Censorship/?f=kids_online.article

Seidensticker, B. (2006). *Futurehype: the myths of technology change.* San Francisco: Berrett–Koehler Publisher, Inc.

Selwyn, N. (2000). Creating a "connected" community? Teachers' use of an electronic discussion group. *Teachers College Record, 102*(4), 750–778.

Senator George J. Mitchell Scholarship Research Institute. (2006). Early college in Maine: Student outcomes and lessons learned from one model. Available online at http://www.mitchellinstitute.org/Gates/pdf/YCCC-WHS_Final_Report.pdf

SENJIT. (2006). Laptops initiative for students with dyslexia and other reading and writing difficulties: Evaluation report 2003–06. Dublin: NCTE. Available online at www.laptopsinitiative.ie

Sharples, M. (2003). Disruptive devices: Mobile technology for conversational learning. *International Journal of Continuing Engineering Education and Lifelong Learning, 12*(5/6), 504–520.

Sheehy, K., Kukulska-Hulme, A., Twining, P., Evans, D., Cook, D., & Jelfs, A., with Ralston, J., Selwood, I., Jones, A., Heppell, S., Scanlon, E., Underwood, J., & McAndrew, P. (2005). Tablet PCs in schools: A review of literature and selected projects, Becta Research. Retrieved November 9, 2006 from http://www.becta.org.uk/corporate/publications/documents/tablet_pc.pdf.pdf

Shields, M. K., & Behrman, R. E. (2000). Children and computer technology: Analysis and recommendations. *Future of Children, 10,* 2. Available online at http://www.futureofchildren.org/usr_doc/vol10no2Art1.pdf

Silvernail, D. L., & Lane, D. M. M. (2004). The impact of Maine's one-to-one laptop program on middle school teachers and students (Report #1). Gorham, ME: Maine Education Policy Research Institute, University of Southern Maine Office. http://mainegov-images.informe.org/mlte/articles/research/MLTIPhaseOneEvaluationReport2004.pdf

Singapore Crescent High School Tablet PCs project: http://www.rsi.sg/english/youngexpressions/view/20040731204630/1/.html

Sloane, F. C. (2005). The scaling of reading interventions: Building multilevel insight. *Reading Research Quarterly, 40,* 3. Retrieved January 21, 2007, from http://www.reading.org/Library/Retrieve.cfm?D=10.1598/RRQ.40.3.4&F=RRQ-40-3-Sloane.html

Stager, G. (2005). Selling the dream of 1:1 computing. Retrieved December 5, 2006, from http://www.stager.org/laptops/talkingpoints/index.html

Stager, G. (2006). *Has educational computing jumped the shark?* Presented at ACEC 2006, Cairns, Australia, October 2, 2006. Retrieved 12/05/2006 from http://www.stager.org/articles/acecshark2006.html

Standage, T. (1998). *The Victorian internet: The remarkable story of the telegraph and the nineteenth centur's on-line pioneers.* New York: Walker Publishing Company.

Stead, G. (2006). Mobile technologies: Transforming the future of learning. In Becta ICT Research, *Emerging technologies for learning.* http://www.becta.org.uk/corporate/publications/documents/Emerging_Technologies.pdf

Stevenson, K. (1999.) Evaluation report—Year 3 middle school laptop program. Retrieved March, 2006, from http://www.beaufort.k12.sc.us/district/evalreport3.htm

Stone B. (2004).The next frontiers: Way cool phones. *Newsweek,* June 7, 2004. http://www.pdatoday.com/more/A1646_0_1_0_M/

Stone, B. (2005). Hi-tech's new day. *Newsweek,* April 11, 2005, 62.

Stufflebeam, D. L. (2000a). The methodology of meta-evaluation as reflected in meta-evaluations by the Western Michigan University Evaluation Center. *Journal of Personnel Evaluation in Education 14*(1), 95–125.

Stufflebeam, D. L. (2000b). The CIPP model for evaluation. In D. L. Stufflebeam, G. F. Madaus, & T. Kellaghan (Eds.), *Evaluation models* (2nd ed.). Boston: Kluwer Academic Publishers.

Stufflebeam, D. L. (2001). The meta-evaluation imperative. *American Journal of Evaluation, 22*(2), 183–209. Retrieved March 18, 2006, from http://aje.sagepub.com/cgi/content/refs/22/2/183

Subrahmanyam, K., Kraut, R., Greenfield, P., & Gross, E. (2000). The impact of home computer use on children's activities and development. *The Future of Children and Computer Technology, 10,* 2, Available online at http://www.futureofchildren.org/usr_doc/vol10no2Art6.pdf

Tapscott, D. (1998). *Growing up digital. The rise of the net generation.* New York: McGraw–Hill.

Tatar, D., Roschelle, J., Vahey, P., & Penuel, W. R. (2003). Handhelds go to school: Lessons learned. *IEEE Computer, 36*(9), 30–37.

Taylor, R. (1980). *The computer in the school: Tutor, tool, tutee.* New York: Teachers College Press.

Taylor, M., & Quayle, E. (2003). *Child pornography: An Internet crime.* London: Routledge.

Technology Counts 05. (2005). Electronic transfer: Moving technology dollars in new directions. *Education Week.* Retrieved October 13, 2005 from http://www.edweek.org/ew/toc/2005/05/05/

Trimmel, M., & Bachmann, J. (2004). Cognitive, social, motivational and health aspects of students in laptop classrooms. *Journal of Computer Assisted Learning, 20*(2), 151–158.

Twining, P., Evans, D., Cook, D., Ralston, J., Selwood, I., Jones, A., et al. (2006). Tablet PCs in schools: Case study report, Becta Research. Retrieved November 9, 2006, from http://www.becta.org.uk/corporate/publications/documents/tabletpc_report.pdf

UNESCO. (1994). The Salamanca statement and framework on special needs education. Paris: UNESCO.

UNESCO. (2003). *Education in and for the information society* (UNESCO publication for the World Summit on the Information Society: Author: Cynthia Guttman). Paris: UNESCO.

University of British Columbia. URL: http://www.sfu.ca/media-lab/

U.S. Census Bureau. (2005). Computer and Internet use in the United States: 2003. U.S. Census Bureau, Current Population Reports 2005. Available online at http://www.census.gov/prod/2005pubs/p23-208.pdf

U.S. Department of Education (1996). *Getting America's students ready for the 21st century: Meeting the technology literacy challenge.* Washington, DC: U.S. Department of Education.

U.S. Department of Education (1997). *Investing in school technology: Strategies to meet the funding challenge.* Washington, DC: U.S. Department of Education.

U.S. Department of Education, National Center for Education Statistics. (2000b). *Teachers' tools for the 21st century: A report on teachers' use of technology.* Washington, DC. Retrieved February 17, 2004, from http://nces.ed.gov/surveys/frss/publications/2000102/

U.S. Department of Education. (2005). Internet access in U.S. public schools and classrooms: 1994–2003, Washington: Institute for Education Science.

U.S. Department of Labor. (1999). Computer ownership up sharply in the 1990s. Summary 99-4, March 1999. Retrieved November 28, 2006, from http://www.bls.gov/opub/ils/pdf/opbils31.pdf

U.S. News & World Report. (April, 2005). In 2005 about 133,000 PCs retired every day in the U.S. alone.

Vahey, P., & Crawford, V. (2002). *Palm™ Education Pioneers Program: Final evaluation report.* Menlo, CA: SRI International. Retrieved March, 18, 2006, from http://www.palmgrants.sri.com/

Valkenburg, P. M., Schouten, A. P., & Peter, J. (2005). Adolescents' identity experiments on the internet, *New Media & Society, 7*(3), 383–402.

Valkenbourg, P., Peters, J., & Schouten, A. (2006). Friend networking sites and their relationship to adolescents' well-being and social self-esteem. *CyberPyschology and Behaviour, 9*(5), 584–590.

Varian, H. R. (2006). A plug for the unplugged $100 laptop computer for developing nations. Available online at www.nytimes.com

Vavoula, G. (2005). Report on literature on mobile learning, science and collaborative activity, *Kaleidoscope Report.* Available online at http://telearn.noekaleidoscope.org/warehouse/vavoula-kaleidoscope-2005.pdf

Venesky, R., & Cassandra, K. (2002). "Quo Vademus?" The transformation of schooling in a networked world. Paris: Organization for Economic Cooperation and Development.

Villani, S. (2001). Impact of media on children and adolescents: A 10-year review of the research. *Journal of the American Academy of Child and Adolescent Psychiatry, 40*(4), 392–401.

Vygotsky, L. S. (1978). *Mind in society: The development of higher psychological processes.* Cambridge, MA: Harvard University Press.

Wang, M., Shen, R., Tong, R., Yang, F., & Han, P. (2005). *Mobile learning with cell phones and Pocket PCs.* Advances in Web-Based Learning—ICWL 2005 4th International Conference, Hong Kong, China, July 31–August 3, 2005. Proceedings.

Ward, M. (2006). Fuel cells to change laptop use BBC news, March 10, 2006. Available online at http://news.bbc.co/uk/2/hi/technology/4794920.stm

Warschauer, M. (2005/2006). Going one-to-one, educational leadership. *Learning in the Digital Age, 63*(4), 34–38.

Warschauer, M. (2006). *Laptops and literacy: Learning in the wireless classroom.* New York: Teacher's College Press.

Warschauer, M., Grant, D., Real, G. D., & Rousseau, M. (2004). Promoting academic literacy with technology: Successful laptop programs in K–12 schools. *System, 32*(14), 525–538.

Wartella, E. A., & Jennings, N. (2000). Children and computers: New technology—Old concerns. *Children and Computer Technology, 10,* 2.

Waycott, J., Jones, A., & Scanlon, E. (2005). PDAs as lifelong learning tools: An activity theory based analysis. *Learning, Media and Technology, 30*(2), 107–130

Weatherley, R., & Lipsky, M. (1977). Street-level bureaucrats and institutional innovation: Implementing special-education reform. *Harvard Educational Review, 47*(2), 171–197, 77.

Weiser, M. (1991). The computer for the 21st century. *Scientific American, 265*(30), 94–104.

Wellman, B. (2002). Little boxes, globalization, and networked individualism. In M. Tanabe, P. van den Besselaar, & T. Ishida (Eds.), *Digital cities: Technologies, experiences, and future perspectives.* Heidelberg: Springer–Verlag.

Wenglinsky, H. (2005). *Using technology wisely: The keys to success in schools,* New York: Teachers College Press.

Wenglinsky, H. (2005/2006). Technology and achievement: The bottom line, *Educational Leadership,* (Special Issue: Learning in the Digital Age), 63, 4.

Winn, M. (2002). *The plug-in drug: Television, computers, and family life* (Rev. ed.). New York: Penguin.

Woodhall, M. (1987). Cost analysis in education. In G. Psacharopoulos (Ed.), *Economics of education: Research and studies* (393–399). New York: Pergamon Press.

Woolcock, M. (1998). Social capital and economic development: Toward a theoretical synthesis and policy framework. *Theory and Society, 27,* 151–208.

Yip, M. (2004) Digitalizing education at Crescent Girls School. *Young Expression,* July 31, 2004. Available online at http://www.rsi.sg/english/youngexpressions/view/20040731204630/1/html

Young, J. R. (2005). MIT researchers unveil a $100 laptop designed to benefit children worldwide. *The Chronicle of Higher Education,* November 25, 2005.

Youra Studio. (2006). Stats on cell phone use. http://www.youra.com/media/images/prsmsstats.pdf

Zhao, Y., & Conway, P. F. (2001). What's in, what's out?: An analysis of state technology plans. *Teachers College Record* [On-line]. Retrieved March, 1, 2006.

Zhao, Y., & Frank, K. (2003). Factors affecting technology uses in schools: An ecological perspective. *American Educational Research Journal, 40*(4), 807–840.

Zhao, Y., Frank, K., & Ellefson, N. (in press). Fostering meaningful teaching and learning with technology: Characteristics of effective professional development efforts. In R. Floden & E. Ashburn (Eds.), *Leadership for meaningful learning with technology.* Teachers College Press.

Zhao, Y., & Lei, J. (forthcoming). New technology. AERA handbook on educational policy research.

Zhao, Y., Lei, J., & Conway, P. (2006). A global perspective on political definitions of e-learning: Commonalities and differences in national educational technology plans. In J. Weiss et al. (Eds.), *The international handbook of virtual learning environments* (pp. 673–697). Dordrecht, the Netherlands: Kluwer/Springer–Verlag.

Zhao, Y., Lei, J., & Frank, K. (2006). The social life of technology: An ecological analysis of technology diffusion in schools. *Pedagogies: An International Journal, 1*(2), 135–149.

Zhao, Y., Pugh, K., Sheldon, S., & Byers, J. L. (2002). Conditions for classroom technology innovations. *Teachers College Record, 104*(3), 482–515.

Zucker, A. A. (2005). *Starting school laptop programs: Lessons learned.* One-to-One Computing Evaluation Consortium: Policy brief number 1. Available online at http://www.ubiqcomputing.org/lit_review.html

Zurita, G., & Nussbaum, M. (2004). A constructivist collaborative learning environment supported by wireless interconnected handhelds. *Journal of Computer Assisted Learning, 20*(4), 235–243.

Zurita, G., Nussbaum, M., & Salinas, R. (2005). Dynamic grouping in collaborative learning supported by wireless handhelds. *Educational Technology and Society, 8*(3), 149–161.

Appendix A: Major ubiquitous computing projects included in this book

---

Canada: Wireless Writing Program
(http://www.prn.bc.ca/Wireless_Writing_Program.html)

Mission-"To improve student achievement, motivation, and learning skills, through the integration of technology with writing instruction"

Start date-September 2003

Scale-17 schools, 1,150 students, and 37 teachers

Device-Laptops

Grade-6–7

Funding source-School district

Research/report-Jeroski (2003, 2005)

Chile: Pocket PCs With Wireless Networks Program

Mission-"To support face-to-face computer supported collaborative learning"

"To change the traditional classroom dynamic from one in which the teacher lectures to passive students to one in which the teacher guides students"

Start date-2004

Scale-8 schools, about 1,900 students

Device-Pocket PCs

Grade-High school

Funding source-Unclear

Research/report-Nussbaum and Zurita (2005)

Ireland: Laptops Initiative
(http://www.ita-kl.de/ita/senistnet/cs12.php)

Mission-"To identify how laptops and other portable ICT equipment can best be used to support students with dyslexia or other reading and writing difficulties in a manner that facilitates learning, and access to learning, in an inclusive environment"

Start date-2000

Scale-31 schools, approximately 1,000 students

Device-Laptops

Grade-Second level

Funding source-Department of Education and Science

Research/report-Conway (2005); Daly (2006)

United Kingdom: Laptops for Teachers Initiative
(http://lft.ngfl.gov.uk/)

Mission-"Using Tablet PCs to support and improve traditional school tasks and using Tablet PCs to extend practice"

Start date-November 2002

Scale-Over 90 schools, many on 1:1 basis

Device-Tablet PCs

Grade-Primary and secondary

Funding source-Various sources

Research/report-Sheehy et al. (2005); Twining et al. (2006)

Singapore: m-learning @ Crescent
(http://www.ida.gov.sg/Infocomm%20Adoption/20061108112645.aspx)

Mission-"To enhance learning and teaching"

Start date-July 2004

Scale-37 teachers and 355 students

Device-Tablet PCs

Grade-Secondary school

Funding source-Buy-in from students and parents

Research/report-Fujitsu (2006)

United Kingdom: the DfES/Becta PDA Project

Mission-"To evaluate initial issues in the use of PDA-type devices in schools both for managing workload and for supporting teaching and learning"

Start date-April 2002

Scale-150 teachers, 100 devices for students in 37 schools

Device-PDAs

Grade-Infant, primary, secondary, and middle schools

Funding source-U.K. Department for Education and Skills

Research/report-Perry (2003)

Hong Kong: Yau Ma Tei Catholic Elementary School

Mission-"Cultivating students' aesthetic sense and vision, developing students' creativity in visual art, and extending the learning space beyond classroom"

Start date-February 2000

Scale-35 laptops for art class

Device-Laptops

Grade-Primary

Funding source-School

Research/report-Hong Kong Case Study Report

United Kingdom, Italy, and Sweden: m-learning Project
(http://www.m-learning.org/index.htm)

Mission-Using mobile technologies to enhance the learning experience.

Start date-2001

Scale-€4.5 million, three countries

Device-Mobile phones, PDAs, pocket PCs, and other mobile devices

Grade-Young adults aged 16–24

Funding source-European Commission's Information Society Technologies (IST) program

Research/report-Attewell (2005)

Maine: Maine Learning Technology Initiative (MLTI)
(http://www.state.me.us/mlte/)

Mission-To prepare Maine's students for a rapidly changing world

Start date-January 2002

Scale-239 schools, 34,000 students, and 3,000 teachers

Device-Laptops

Grade-7–8

Funding source-State

Research/report-Lemke and Martin (2004); Silvernail and Harris (2003); Silvernail and Lane (2004)

Michigan: Freedom to Learn (FTL) Project
(http://wireless.mivu.org/index.cfm)

Mission-Improving student achievement and engagement

Start date-Initiated in 2001 and expanded in 2003

Scale-Nearly 30,000 students in 15 school districts

Device-Laptops, PDAs

Grade-6

Funding source-State

Research/report-Urban-Lurain and Zhao (2004, 2005)

New Hampshire: Technology Promoting Student Excellence (TPSE) Project
(http://www.bc.edu/research/intasc/studies/nhLaptop/description.shtml)

Mission-Expanding learning opportunities and erasing the digital divide

Start date-2004

Scale-Six neediest schools, over 400 students, and 35 teachers

Device-Laptops

Grade-7

Funding source-State (U.S.$1.2 million raised from 24 private organizations)

Research/report-Bebell (2005)

Virginia: Henrico County Public Schools (HCPS) Project

Mission-Providing students with a quality education and teaching them state-of-the-art technology skills to prepare them for college and the workplace

Start date-2001

Scale-24,000 laptops to all students, 3,300 laptops to entire teaching and administrative staff

Device-Laptops

Grade-6–12

Funding source-School district

Research/report-Davis, Garas, Hopstock, Kellum, and Stephenson (2005); Zucker and McGhee (2005)

Pennsylvania: Quaker Valley Digital School District
(http://www.qvsd.org/2550115914103416/site/default.asp)

Mission-"Bridging the digital divide"

To understand "how the information revolution can be harnessed to create an education revolution"

Start date-2001

Device-Laptops

Scale-About 2000

Grade-Started with grades 3–12, but then changed to grades 9–12 (2004–2005 academic year)

Funding source-State

Research/report-Faulk (2003); Kerr, Pane, and Barney (2003)

United States: Alpha Middle School 1-1 Laptop Project

Mission-Preparing students for what will be the future of teaching and learning

Start date-2003

Scale-About 250 students and 40 teachers

Grade-7–8

Device-Laptops

Funding source-School

Research/report-Lei (2005); Lei and Zhao (in press)

New York: Rye Country Day School project
(http://www.netdaycompass.org/outside_frame.cfm?thispath=instance_
id=2454^category_id=5&thislink=http://www.rcds.rye.ny.us/rcds_laptop_program/&instance_id=3200)

Mission-To use advancing computer technology to improve the education process

Start date-1999

Scale-About 500 students

Device-Laptops

Grade-Started with 7–10, expanded to 7–12

Funding source-Students

Research/report-Bartels, F. (2000, 2002)

United States: The Palm™ Education Pioneer (PEP) Program

Mission-To evaluate the potential of handheld computers for K–12 teaching and learning

Start date-2001

Scale-175 classrooms in the United States

Device-PDAs

Grade-K–12 ˋ

Funding source-Palm, Inc.

Research/report-Tatar, Roschelle, Vahey, and Penuel (2003); Vahey and Crawford (2002)

Appendix B:
Studies included
in Penuel's (2005)
synthesis of research on
one-to-one initiatives

Implementation studies

Daitzman, P. (2003). *Evaluation of the national model laptop program technology literacy: A journey into the global learning community.* New Haven, CT: East Rock Global Magnet School.

Davies, A. (2004). *Finding proof of learning in a one-to-one computing classroom.* Courtenay, B.C.: Connections Publishing.

Davis, D., Garas, N., Hopstock, P., Kellum, A., & Stephenson, T. (2005). Henrico County Public Schools iBook survey report. Arlington, VA: Development Associates, Inc.

Dinnocenti, S. T. (2002). *Laptop computers in an elementary school: Perspectives on learning environments from students, teachers, administrators, and parents.* Unpublished doctoral dissertation, University of Connecticut, Storrs, CT.

Fairman, J. (2004). *Trading roles: Teachers and students learn with technology.* Orono, ME: Maine Education Policy Research Institute, University of Maine Office.

Gaynor, I. W., & Fraser, B. J. (2003). *Online collaborative projects: A journey for two year-5 technology rich classrooms.* Paper presented at the Western Australian Institute for Educational Research Forum.

Harris, W. J., & Smith, L. (2004). *Laptop use by seventh grade students with disabilities: Perceptions of special education teachers.* Orono, ME: Maine Education Policy Research Institute, University of Maine Office.

Hill, J., & Reeves, T. (2004). *Change takes time: The promise of ubiquitous computing in schools. A report of a four-year evaluation of the laptop initiative at Athens Academy.* Athens, GA: University of Georgia.

Jaillet, A. (2004). What is happening with portable computers in schools? *Journal of Science Education and Technology, 13*(1), 115–128.

Jeroski, S. (2003). *Wireless Writing Project School District No. 60 (Peace River North) research report: Phase II.* Vancouver, BC: Horizon Research and Evaluation, Inc.

Kemker, K., & Barron, A. (2004). *Laptop computers as a tool for authentic instruction.* Paper presented at the Annual Meeting of the National Educational Computing Conference, New Orleans, LA.

Kerr, K. A., Pane, J. F., & Barney, H. (2003). *Quaker Valley Digital School District: Early effects and plans for future evaluation.* Santa Monica, CA: RAND.

Lane, D. M. M. (2003). *The Maine Learning Technology Initiative impact on students and learning.* Portland, ME: Center for Education Policy, Applied Research, and Evaluation, University of Southern Maine.

Light, D., McDermott, M., & Honey, M. (2002). *Project Hiller: The impact of ubiquitous portable technology on an urban school.* New York: Center for Children and Technology, Education Development Center.

Lowther, D. L., & Ross, S. M. (2003, April). *When each one has one: The influences on teaching strategies and student achievement of using laptops in the classroom.* Paper presented at the Annual Meeting of the American Educational Research Association, Chicago, IL.

Lowther, D. L., Ross, S. M., & Morrison, G. R. (2001, July). *Evaluation of a laptop program: Successes and recommendations.* Paper presented at the National Education Computing Conference, Chicago, IL.

Mitchell Institute. (2004). *One-to-one laptops in a high school environment: Piscataquis Community High School Study final report.* Portland, ME: Great Maine Schools Project, George J. Mitchell Scholarship Research Institute.

Newhouse, C. P. (2001). A follow-up study of students using portable computers at a secondary school. *British Journal of Educational Technology, 32*(2), 209–219.

Newhouse, C. P., & Rennie, L. (2001). A longitudinal study of the use of student-owned portable computers in a secondary school. *Computers and Education, 36*(3), 223–243.

Silvernail, D., & Lane, D. M. M. (2004). *The impact of Maine's one-to-one laptop program on middle school teachers and students: Phase one summary evidence.* Portland, ME: Maine Education Policy Research Institute, University of Southern Maine.

Silvernail, D. L., & Harris, W. J. (2003). *The Maine Learning Technology Initiative teacher, student, and school perspectives: Mid-year evaluation report.* Portland, ME: Maine Education Policy Research Institute, University of Southern Maine.

Stevenson, K. R. (2002). *Evaluation report—Year 2: High school laptop computer program (final report, for school year 2001–2002).* Liverpool: Liverpool Central School District, New York.

Texas Education Agency. (2001). *Report on the Ed Tech PILOTS.* Austin: Texas Education Agency.

Trimmel, M., & Bachmann, J. (2004). Cognitive, social, motivational and health aspects of students in laptop classrooms. *Journal of Computer-Assisted Learning, 20*(2), 151–158.

Warschauer, M., Grant, D., Real, G. D., & Rousseau, M. (2004). Promoting academic literacy with technology: Successful laptop programs in K–12 schools. *System, 32*(14), 525–538.

Windschitl, M., & Sahl, K. (2002). Tracing teachers' use of technology in a laptop computer school: The interplay of teacher beliefs, social dynamics, and institutional culture. *American Educational Research Journal, 39*(1), 165–205.

Zucker, A. A., & McGhee, R. (2005). *A study of one-to-one computer use in mathematics and science instruction at the secondary level in Henrico County Public Schools.* Arlington, VA: SRI International.

Outcome studies

Gulek, J. C., & Demirtas, H. (2005). Learning with technology: The impact of laptop use on student achievement. *Journal of Technology, Learning, and Assessment, 3*, 2.

Russell, M., Bebell, D., & Higgins, J. (2004). *Laptop learning: A comparison of teaching and learning in upper elementary classrooms equipped with shared carts of laptops and permanent 1:1 laptops.* Boston: Technology and Assessment Study Collaborative, Boston College.

Schaumburg, H. (2001, June). *Fostering girls' computer literacy through laptop learning: Can mobile computers help to level out the gender difference?* Paper presented at the National Educational Computing Conference, Chicago, IL.

Index